Robotic Sailing 2016

José C. Alves · Nuno A. Cruz
Editors

Robotic Sailing 2016

Proceedings of the 9th International Robotic Sailing Conference

Springer

Editors
José C. Alves
Department of Electrical and Computer Engineering
University of Porto
Porto
Portugal

and

INESC Technology and Science
Centre for Robotics and Autonomous Systems
Porto
Portugal

Nuno A. Cruz
Department of Electrical and Computer Engineering
University of Porto
Porto
Portugal

and

INESC Technology and Science
Centre for Robotics and Autonomous Systems
Porto
Portugal

ISBN 978-3-319-45452-8 ISBN 978-3-319-45453-5 (eBook)
DOI 10.1007/978-3-319-45453-5

Library of Congress Control Number: 2016949115

Printed on acid-free paper

This Springer imprint is published by Springer Nature
The registered company is Springer International Publishing AG
The registered company address is: Gewerbestrasse 11, 6330 Cham, Switzerland

Preface

Humans sail the oceans for centuries. In our highly technological age, a new interest on ancient forms of navigation is gaining a new attention with the necessity to know more and more about the oceans that undoubtedly regulate the delicate ecosystem where we live. Persistent ocean monitoring is a necessity and is currently being undertaken by a multitude of robotic platforms throughout the globe. Unmanned wind propelled crafts have a great potential for providing ocean data, with superior performance than other robotic technologies in terms of autonomy, speed and maneuverability, and also the ability to provide a continuous remote access for data retrieval and mission control.

Sailing is a challenging task driven by highly complex dynamic interactions between the wind, the water and all the sailboat's components above and below the surface. The design, construction and navigation of a sailing boat is the result of a multifaceted combination of several engineering fields, with continuous progresses towards more efficient ways to use the wind power while withstanding the harsh marine environment. Removing the crew and automating the navigation of a sailing boat raises this level of complexity while transferring to a computer system all the knowledge, experience and intuition that makes a human sailor handling correctly a sailing boat.

In the last few years there have been various developments in this field and some successful autonomous sailing boats have already performed relevant missions in the oceans. About ten years ago, a small community engaged in this domain has been challenged to develop an autonomous sailing boat to cross the Atlantic without assistance, resembling the first defiance in the early 60s for solitaire Atlantic crossing. Although not yet succeeded, the Microtransat challenge has stimulated various researchers from universities and companies, and also enthusiasts of sailing, to start working towards this objective. The World Robotic Sailing Championship and the International Robotic Sailing Conference was born in 2008 to provide an annual scientific and experimental forum to join the world community, share results and demonstrate their last achievements in the field.

The 9th World Robotic Sailing Championship and International Robotic Sailing Conference returns to Portugal after the successful WRSC/IRSC 2009 organized in

Matosinhos by the Faculty of Engineering of the University of Porto. The 2016 edition will be once again organized by the Faculty of Engineering of the University of Porto (FEUP), now in cooperation with the Institute for Systems and Computer Engineering, Technology and Science (INESC TEC), and hosted by the city of Viana do Castelo, north of Portugal. This is a city with a strong tradition in sailing and a recognized engagement in the promotion of water sports in the region.

These proceedings contain the papers selected for presentation in the 9th International Robotic Sailing Conference and are organized in three main topics. The first part addresses the design, construction and validation of new platforms and rigs. The second part is devoted to the development of sensors and algorithms to enhance the performance of critical maneuvers of robotic sailing boats. Finally, the papers in the last part are dedicated to the improvement of behaviors required for autonomous missions.

We would like to acknowledge the effort of those who contributed to the organization of WRSC/IRSC 2016. This includes not only the authors of the papers and the participants in the competition, but also the members of the Program Committee for their valuable and timely reviews that greatly improved the overall quality of the papers. We also would like to thank the support of the organizing entities (Faculty of Engineering of the University of Porto and the Institute for Systems and Computer Engineering, Technology and Science), the Municipality of Viana do Castelo and the Sailing Club of Viana do Castelo for hosting this event in the city, and all the other sponsors, colleagues and friends that also contributed to the success of WRSC/IRSC 2016.

Porto, Portugal
July 2016

José C. Alves
Nuno A. Cruz

Organizing Committees

General Chairs

José C. Alves, Universidade do Porto/INESC TEC, Portugal
Nuno A. Cruz, Universidade do Porto/INESC TEC, Portugal

Program Committee

Fabrice Le Bars, École Nationale Supérieure de Techniques Avancées (ENSTA) Bretagne, France
Taylor Barton, University of Texas, Dallas, US
Ole Blaurock, Fachhochschule Lübeck, Germany
Benoit Clement, École Nationale Supérieure de Techniques Avancées (ENSTA) Bretagne, France
Anna Friebe, Åland University of Applied Sciences, Finland
Oren Gal, Technion Israel Institute of Technology, Israel
Jorge Cabrera Gamez, Universidad de Las Palmas de Gran Canaria, Spain
Florian Haug, Åland University of Applied Sciences, Finland
Henrik Hesse, ETH, Automatic Control Laboratory, Switzerland
Luc Jaulin, École Nationale Supérieure de Techniques Avancées (ENSTA) Bretagne, France
Hugo Kerhascoët, NKE Marine Electronics, France
Erik Maehle, University of Lübeck, Germany
Benedita Malheiro, Instituto Superior de Engenharia do Porto, Portugal
Paul Miller, US Naval Academy, USA
Cedric Pradalier, GeorgiaTech Lorraine, France
Kostia Roncin, École Nationale Supérieure de Techniques Avancées (ENSTA) Bretagne, France
Colin Sauze, Aberystwyth University, Wales, UK

Michael Schukat, National University of Ireland, Galway, Ireland
Alexander Schlaefer, Hamburg University of Technology, Germany
Roland Stelzer, Happylab GmbH, Austria
Dermot Tynan, Hewlett Packard, Ireland
Diedrich Wolter, University of Bremen, Germany

Contents

Part I
Sailboat Platforms

Free Rotating Wingsail Arrangement for Åland Sailing Robots

Teo Enqvist, Anna Friebe and Florian Haug

Abstract This paper accounts for our ongoing work to evaluate, design and build an optimized wingsail arrangement for one of Åland Sailing Robots' boats. By examining the current conventional sails of the boat and its propulsion capability, a wingsail arrangement with a similar propulsion capability has been designed and further investigated. The final design has to meet several criteria such as simplicity, reliability, and high degree of autonomous operation. A thorough analysis of the design criteria resulted in the decision to choose a symmetrical, free-rotating wingsail with an additional tail for actuating the wing and controlling its angle of attack. The chosen design is further investigated by using a CFD simulation software. Furthermore, we discuss possible solutions and recommendations regarding the future physical construction of the wingsail.

1 Introduction

A rigid wingsail is an interesting option for robotic sailboats, with many advantages, as will be described in this paper. In Åland Sailing Robots this alternative is evaluated as a rig for a Mini 12 (2.4 mR class) sailboat hull, see Fig. 1.

Åland Sailing Robots (ÅSR) was initiated at the Åland University of Applied Sciences (Åland UAS) in 2013, with the goal of a transatlantic sailing with an autonomous sailboat. Starting with rebuilding a radio controlled 1 m sailboat and transitioning to a 4 m Mini 12, the project has proceeded and contributed research on power management strategies [5], modelling, control and state-estimation [12], as well as a review of challenges for autonomous sailboats on long journeys [9].

T. Enqvist (✉) · A. Friebe · F. Haug
Åland University of Applied Sciences, PB 1010, 22111 Mariehamn, Åland, Finland
e-mail: teo.enqvist@ha.ax

A. Friebe
e-mail: anna.friebe@ha.ax

F. Haug
e-mail: florian.haug@ha.ax

J.C. Alves and N.A. Cruz (eds.), *Robotic Sailing 2016*,
DOI 10.1007/978-3-319-45453-5_1

Fig. 1 Åland Sailing Robots Mini 12 sailboat, which shall be fitted with a wingsail

In 2016–2018 a project is running at the Åland UAS with the goal to develop a mobile marine research platform that is autonomous and utilising wind and solar energy. This project, Marine Research Platform - Åland Sailing Robots is partly funded by the European Regional Development Fund. The free rotating wingsail described in this paper is one of the considered options for the marine research platform rig.

The Mini 12 is a one-manned sailboat models with a fixed keel and a waterline under 4 m. Mini 12 is a synonym for the competitive international 2.4 Metre Class. The name Mini 12 for the boats comes from the first boats constructed for the class in a scale of 1:5 of the original 12 m yachts (12/5 = 2.4) [21]. The calculated sail area for the SR Mini 12 is 8.2 m^2. The largest combined area of the main and head sails available for the Mini 12 is 11.7 m^2 (which is a size deemed ineffective in most wind situations due to the large size of the sails). The mass of the Mini 12 is approximated as 300 kg [11]. These parameters are used when calculating the lift and drag forces the sails of the Mini 12 create.

Wingsails are solid symmetrical wings vertically fitted on the vessel. These shouldn't be confused with solid square sails or rigid sails. Cloth sails must be furled when not in use in order to prevent destructive flogging, while the wing can be oriented directly to the wind in a way so that it experiences a minimal aerodynamic force. Wingsails have almost exclusively symmetrical airfoil sections because of the need to be operated with the wind blowing on either side of the ship. Symmetrical solid airfoil sections cannot create particularly high lift coefficients by themselves. Therefore most wingsail arrangements include a flap or tail section which creates lift and helps the main wing to reach its maximum lift capacity. The advantages of wingsails have been noted and suggestions for use of wingsails for fuel saving in commercial ships [19] as well as for autonomous sailing have been made. Several examples of rigid wingsails in autonomous sailboats and hybrid boats have been described previously [1, 3, 4, 6–8, 15–17]. These have been applied to boats ranging in size from 72 cm [17] to nearly 6 m [7], and for monohulls [6, 15, 16] as well as catamarans [4, 7, 8].

2 State of the Art

The by far most common setup for sailing a boat is a soft (cloth) sail which is manually trimmed in order to generate an optimal lift-to-drag ratio given a predestined course of the boat. However, considering autonomous sailing, the actuation of the sail is non-trivial and the forces required to hold the sail in position are quite large. Forces can be decreased by using a balanced rig design, and in theory this could lead to savings of two-thirds of the sail trim power [20]. In practice, however, it has proven difficult to obtain those large savings [13]. Moreover, the lift-to-drag ratio for conventional sails is about 3–5, which is relatively low [8].

On the other hand, rigid wingsail arrangements have been used for more than 100 years in aviation and become more and more common even in sailing applications. A much higher lift-to-drag ratio in the range of around 100 below stall [19] allows using considerably smaller sail areas. This property together with the fact that a wing sail has no boom, reduces the forces necessary to hold the wing fixed in a certain position considerably. Moreover, by mounting an extra tail behind the main wingsail, the wingsail arrangement can be made self-trimming. The self-trimming capability is achieved by aligning the center of mass of the wing arrangement at the aerodynamic center of the main wing. This way the wing arrangement will work much like a wind vane meaning it will turn in to the wind. The self-trimming ability will be discussed in detail later in this text.

The rigid form of the wings makes it easier and more predictable than soft sails in different wind scenarios. This is because a rigid wing doesn't suffer from aeroelastic collapse (also known as luffing) when pointed high into the wind, unlike soft sails. Aeroelastic collapse causes a great deal of drag and limits the angle the boat can sail into the wind. The rigid wing does not suffer from any aeroelastic problems and can point straight into the wind with very little drag. as will be shown in Sect. 4.2, Fig. 3. Mounting the wingsail free-rotating around its aerodynamic center and twisting the tail in relation to the main wing, allows to control wingsail arrangement entirely. Control of the tail angle allows therefore to adjust an arbitrary angle of attack for the main sail.

Thus, the simplicity and the efficiency of such a wingsail application promise to be ideal for an autonomous sailing robot, both with respect to the boats sailing efficiency and its energy efficiency. The free rotating/self trimming ability of the tailed wing are relatively easy to control; only a small actuator is needed to adjust the tail. At the same time, the wing sections are efficient in creating lift (thrust). The method of controlling the wing arrangement will have a great influence in choosing the wing most suitable for ÅSR Mini 12. A possible future steering method will be discussed more in detail in Sect. 6.

3 Theory Behind the Wing

In many aspects, wingsails and aircraft wings behave rather similar and have common design concepts. For example, the wings should be both light and sturdy at the same time. However, there exist also fundamental differences between wingsails mounted on sailboats and wings mounted on aircraft. The most obvious difference is that a wingsail has to be symmetric. The reason is that a sailboat must be able to sail with the wind coming from both port and starboard, whilst most aircrafts only fly upright. Since symmetric airfoil designs have a lower maximum lift coefficient it is important to optimize the wingsails profile in order to achieve as high lift coefficients and lift-to-drag ratios as possible.

Another important aspect when designing a wingsail is the large difference in operation conditions between wingsails and aircraft wings. Aircrafts operate at high altitudes in mostly lower temperatures and at relatively high speeds, whereas sailboats operate at sea level, often at higher temperatures and at low speeds. Low air speeds are synonymous to low Reynolds numbers. It is known that low Reynolds numbers have a large influence on a rigid wingsail and can lead to unwanted effects such as laminar separation bubbles or stalling. These effects will be described in more detail in the next subsection.

3.1 Reynolds Number Effects

The Reynolds number is a dimensionless quantity that represents the ratio between kinematic or inertial forces to the viscous forces in a fluid, meaning the ratio of force required to push the fluid out of the way versus the force required to slip through the fluid [8]. The Reynolds number R_e is used in airfoil design to manage "scale effects" when computing or comparing wing characteristics, since a small wing will perform differently if the wing is scaled up and vice versa [2].

Neglecting Reynolds number effects has been one of the largest reasons for wing-sail applications failing to break through despite the superior efficiency of wingsails compared to conventional soft sails. Neglecting the Reynolds number leads to poor performance and that in turn, has delayed the transition of soft rigs to rigid wings on sailboats [8].

Low Reynolds numbers are usually connected to small model airplanes (i.e. gliders) and they share similarities with wing sails. Firstly, both the model airplanes and the sailboat require a high lift to drag ratio. For a sailboat this leads to the ability to point upwind, whilst a glider has the ability to glide through the air without sinking rapidly downwards. Also, both the model glider and sailboat require a high lift coefficient. This leads to an increased sailing speed for the boat and, in the case of gliders, the ability of slow flight when tightly circling upward currents of warm air (thermals) [8].

For low Reynolds numbers, however, airfoil sections design suffer from several effects making high lift sections design difficult. Selig, M.D. et al. has written an extensive monograph titled "Airfoils at low speeds" which describes in detail the effects of low Reynolds numbers for airfoils [18]. Most important for sailboats are two effects called *laminar separation bubbles* and *stalling*, which will be described in the following.

3.1.1 Laminar Separation Bubbles

The flow behaviour over an airfoil at high Reynolds numbers (in the millions) is well known. From the leading edge to around the mid-chord, the boundary layer is laminar and here the flow makes a transition to turbulent flow. The transition and the turbulent flow behind it are generally well behaved.

At Reynolds numbers from 50,000–500,000 (low Reynolds number regime) the flow characteristics are fundamentally different and more complicated than at high Reynolds numbers. The transition process is not abrupt, nor does it usually take place while the boundary layer is attached to the airfoil. Instead, the laminar boundary layer separates, i.e. it physically detaches from the surface of the airfoil. While separated, the flow becomes unstable and makes the transition to turbulent flow "mid-air", and only then the flow reattaches to the airfoil. Sometimes, if the Reynolds number is very low or the laminar separation point is sufficiently far aft, the flow entirely fails to return to the airfoil surface, leading to large energy losses associated with this process. This laminar separation, transition to turbulence, and turbulent reattachment enclose a region of recirculating flow is called the "laminar separation bubble". The bubble increases drag and reduces lift depending on the bubbles length because of the varying pressure distribution over the airfoil caused by the bubble. The presence and behaviour of the laminar separation bubble is dependent upon the Reynolds number, airfoil shape, free stream turbulence, surface roughness and sound waves. Efforts towards drag reduction largely concentrate on reducing the size and extent of the laminar separation bubble [18].

3.1.2 Stalling

In aviation, stalling usually refers to the loss of lifting forces and the increasing of drag forces acting over the wings due to an angle exceeding the critical angle of attack.

Flow separation from an airfoil cause stalling, which leads to loss of lift and a rise in drag forces over an airfoil. Stalling occurs when the wing exceeds the critical angle of attack and starts to lose lift and increase its drag. Usually, one distinguishes between two kind of stalling: *Leading edge stalling* occurs when a laminar separation bubble near the leading edge of an airfoil shortens and suddenly bursts. At low angles of attack a laminar flow region separates and reattaches forming a bubble and as the angle increases, the bubble shortens and moves forward and at some point the boundary layer can't reattach and the bubble bursts. When the bubble bursts the wing suddenly loses its lifting forces and stalls. This type of stall is exhibited by airfoils with a thickness between 9 and 15 % of the chord length, and at low Reynolds numbers [2]. On the other hand, *trailing edge stalling* occurs when a turbulent boundary layer at the rear of an airfoil separates. The point where separation occurs transits forward as the angle of attack increases. This type of stall is exhibited by many airfoil sections with a thickness of greater than 15 % of the chord length and even at higher Reynolds numbers [2].

4 Design Criteria for Wingsail

Even though the soft sails fitted on the Mini 12 are proven functional, the search for a more efficient robotic sailboat continues. The transition from a soft sail to a wingsail is an important and promising step to increase the efficiency of Åland Sailing Robots Mini 12. Efficiency is one of the most important factors when discussing robotic sailboats, but also other factors play a part, such as:

1. Economic viability
2. Simplicity
3. Reliability
4. Autonomous operation
5. Design and installation

In the following, we will in short discuss the five factors and how they affect the wing sail application for ÅSR Mini 12.

Economic viability: In this first step of an ongoing project, solely a theoretical analysis of the possibility of fitting a wingsail on the Mini 12 has been performed. Depending on the results of this theoretical investigation, a decision will be made regarding the building and fitting of the wing. Just like this project is made by a student at Åland UAS, most likely so will the building and fitting of the wing be too. The costs of construction and fitting of the wing will be relatively low because of the fact that it will be mostly student working on the wing, and doing maintenance on it. This puts close to all the costs solely on materials and components.

Simplicity As described in Sect. 2 above, the wingsail is a relatively "simple" construction in itself. It is a free rotating symmetrical airfoil with a controllable tail section consisting of a smaller symmetrical airfoil. The fact that the wing is free rotating (self-trimming) makes controlling of the wing simple because the only component that requires control is the tail section while the main wing is steered by the wind itself. This is an advantage over the soft sail application that the Mini 12 is fitted with concerning energy consumption for steering the sail.

Reliability A final evaluation of the reliability can first be made when the wingsail is constructed, mounted and tested together with all peripheral components. However, the fact that only the tail section is to be controllable makes steering the wing application "relatively simple", meaning that in order to steer the wingsail application not too many components are needed. Less components makes the risk for failure smaller, but it also makes recovery from one component failing much harder which means that some effort has to be spent to build a redundant system. However, in summary it seems to be unlikely that the reliability of the wing will be lessened compared to the soft sails.

Autonomous operationality The free rotating and self-trimming ability of the wing arrangement makes it ideal for autonomous sailing. The wing doesn't need very powerful and fast acting servos which constantly re-trim the sails, but uses the wind itself to steer the main wing. The manoeuvres used with the wing sail arrangement are gentle and controlled because of the free rotating capability which makes only the tail section requiring some sort of control. Combining the self-trimming ability and the efficiency to create thrust, the self-trimming wing arrangement is very suitable for autonomous operationality.

Design and installation The installation of a wingsail should be possible without major difficulty. This can be an issue because of the size the wing must be adopted to the size of the Mini 12. A wing is a lot more difficult to get on/off the mast than for example raising/lowering soft sails. Installation may prove to be the a disadvantage of wingsails over soft sails, depending on how it will be mounted on the mast. The rigid form of the wing makes assembly/disassembly by hand difficult and in worst case a sort of crane must be used in order to mount the wing to the mast. Some problems can occur if the boat is moored for a longer time and the wing is not disassembled. However, having the wing freely rotating and positioned in a neutral setting allows the wing arrangement to freely rotate into the wind without creating thrust. This requires, however, the boat to be positioned in a way that the turning radius is at safe distance to anything the wing could hit while turning freely.

The weight of the wing arrangement should be as low as possible, not only to keep the mass center of the boat low but also to ease installation. Design aspects of the wing arrangement will be discussed more in detail later in this text.

4.1 A Self-trimming Wing Arrangement

As stated earlier several times, a self-trimming wing arrangement works basically as a wind vane, where the wind itself steers the wing to point directly into the wind and thus eliminating the need to constantly steer the wing in minor variations in wind direction. When fitted with a tail to steer the main wing the "wing arrangement" can maintain maximum efficiency in any wind direction without disturbance. Also the wing arrangement doesn't cause the sailboat to heel as much as a cloth sail in different wind directions because the wing turns straight in to the wind. This free rotating capability is achieved by aligning the aerodynamic center of the main wing and the mass center of the wing arrangement at the mast where bearings are fitted which allow the wing to rotate with close to zero resistance. The mass center must be manually aligned using ballast in front of the wing in order to get it to align with the mast. Some ideas about how the need for ballast can be exploited in a favourable way will be presented in Sect. 6.

The aerodynamic center of a low speed airfoil is positioned approximately one quarter of the chord behind the leading edge [14] and characterizes the position where the magnitude of the aerodynamic moment remains nearly constant for all angles of attack. For symmetrical airfoils, the aerodynamic moment about the aerodynamic center is zero for all angels of attack. By positioning the mast at the aerodynamic center, the pitching moment for the main wing is zero. In this situation, the tail can be used to force the main wing to adjust at a certain angle of attack thereby generating thrust to the boat. Figure 2 shows in a simplified way the self-trimming capacity of the wing arrangement and depicts how tacks and jibes can be performed. When the tail is aligned perfectly along the centerline of the main wing (Fig. 2b, e), there is no

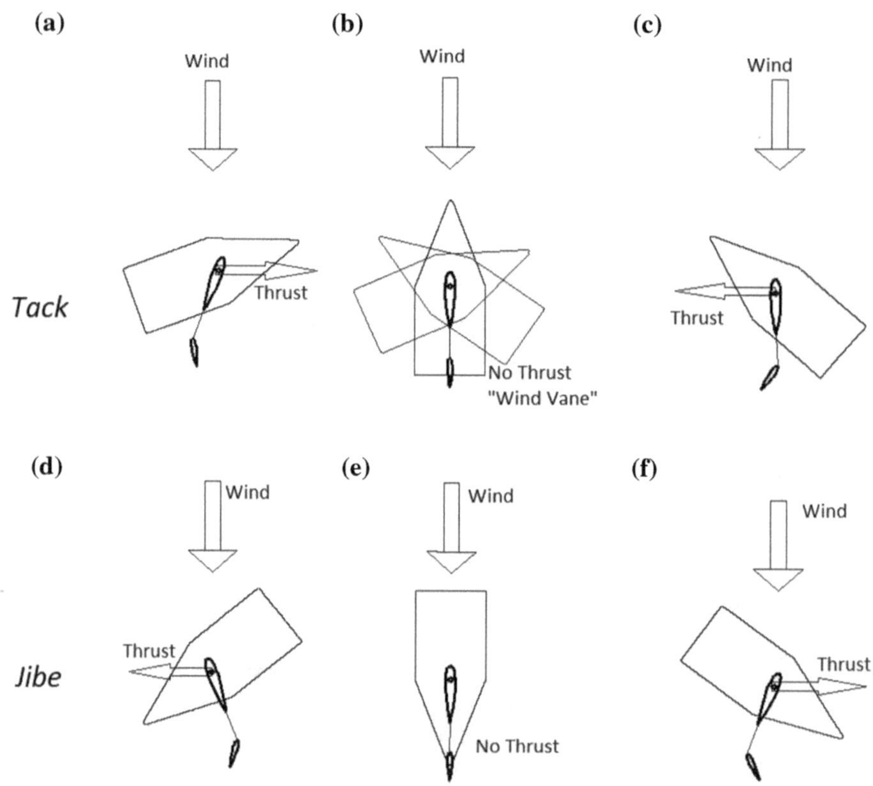

Fig. 2 Simplified sketch of the concept behind a self-trimming wing arrangement. Note the tail section asserts a stabilizing force to keep the main wing at a desirable angle to the wind. *The upper line* (**a–c**) shows a typical tack maneuver, whereas the *lower line* (**d–f**) depict a jibe maneuver. See text for more details

angle of attack and the sail arrangement acts as a weather vane, pointing exactly into the relative wind. In this situation, no thrust is generated. However, by rotating the tail either anticlockwise (Fig. 2a) or clockwise (Fig. 2b), a moment is created and the main wing will parry in the opposite direction in order to balance the moment. This, in turn, will generate thrust which can propel the sailboat as indicated in the figure. When the wind changes in strength or velocity, the wing arrangement will self-adjust accordingly and the course can be kept. In order to tack the boat (or to tack the sail, in case the wind would cross the centerline of the boat), the angle of attack of the tail has to be changed from negative to positive (or vice versa). The main wing will adopt to the new situation, and the thrust is now in the opposite direction, see Fig. 2a–c. In a similar way, a jibe maneuver can be performed, as depicted in Fig. 2d, e. Note that both maneuvers will happen very smooth, without i.e. without sail luffing, violent sail/boom swinging or large change in heeling angle.

4.2 *Design of the Wingsail*

In order to gain a more exact picture of the involved forces and momenta, and for being able to perform computational flow dynamics (CFD) simulations, some theoretical background is needed. For our purpose, simplified lifting theory can be used, which is the same for sailboats as for planes.

According to simplified lifting theory, the lift and drag forces L and D satisfy the equations

$$L = \frac{1}{2}\rho \cdot A \cdot C_L \cdot v^2 \qquad (1)$$
$$D = \frac{1}{2}\rho \cdot A \cdot C_D \cdot v^2$$

where C_L and C_D is respectively the Lift and Drag coefficient, ρ is the air density, v is the true airspeed and A is the sail area.

Knowing the lift and drag coefficients for different sails or wing profiles allows us to calculate the lift and drag forces the sail or wing profile creates. Conventional sails can generate maximum lift coefficients of about 0.8, if they are perfectly trimmed.[1] In order to design a wing with similar or better lift capability we use therefore $C_L = 0.8$ for our calculations and simulations. Estimating the drag coefficient of a conventional sail is more difficult, and our knowledge about a realistic coefficient is limited. However, since most airfoils have a drag coefficient smaller than a bare

[1]To our knowledge, the theoretical prediction of $C_{L_{max}}$ is still a very hot topic in sophisticated CFD research, and while there are promising results there is not yet a general model which works like the "real world". Our estimated value $C_{L_{max}} = 0,8$ is based on [8]. Note, however, that this value is only indicative and will guide us to design a first prototype of a wing sail. The wing sails performance in comparison to the soft sail's has then to be a measure for the reliability of our theoretical assumptions.

mast [8], it is safe to assume that the drag will decrease immensely if the cloth sail is exchanged to a rigid wing sail.

The sail area A_s of a conventional soft sail can be calculated according to

$$A_s \approx P \cdot E \cdot k \quad (2)$$

where P is the height of the sail, E is the width of the sail and k is chosen as 0.62 when calculating the main sail area and 0.5 when calculating the area of the jibs.

Likewise, the area A_w of a wingsail is given by the equation

$$A_w = c \cdot h \quad (3)$$

with c being the length of the chord and h the height of the wing. Note that the wing area is calculated as the planform area (as seen from above) looking along the "lift" direction and *not* the as whole wing surface area (NASA, 2016).

The current sail area for the Mini 12 is calculated to be 8.2 m^2. By using Eq. 2 the sails lift force can be estimated to be maximal about 100 N at a wind speed of 5 m/s.

For the main wingsail, we selected a NACA0021 profile with a chord length of 1 m and a height of 3.5 m. The NACA0021 is chosen because of its good lift and drag properties at quite large span of angles of attack, see Fig. 3.

The 1 m wing has a characteristic length of 1.05 m, resulting in a Reynolds number of about 350 000 at the same wind velocity of 5 m/s and an air density of 1.204 kg/m^3. The corresponding maximum lift coefficient $C_L = 1.732$ was retrieved with the help of Javafoil which gives us a calculated maximal lift force of 91.2 N and a drag force of approximately 13 N at 21° incidence (see Fig. 3, left). This wing creates almost as much thrust as the generously estimated thrust of the soft sails.

Note however, that Javafoil does not take in account the laminar separation bubble or flow separation and thus stalling characteristics cannot be calculated. [10].

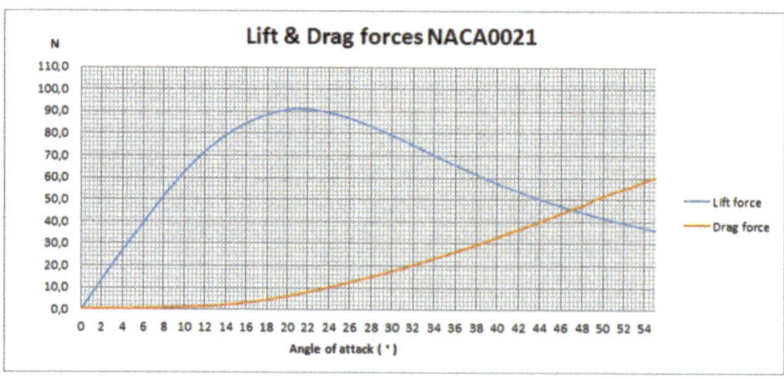

m/s	Clmax	Re
1	1.644	70 000
2	1.679	140 000
3	1.704	210 000
4	1.721	280 000
5	1.732	350 000
6	1.741	420 000
7	1.748	490 000
8	1.753	555 000
9	1.758	625 000
10	1.763	695 000
11	1.766	765 000
12	1.769	835 000
13	1.772	905 000
14	1.775	975 000
15	1.778	1 050 000

Fig. 3 *Left* Lift and Drag forces calculated using the Eq. 2. The lift and drag coefficients are retrieved from Javafoil for a NACA0021 airfoil at Re 350000. *Right* Calculated Reynolds number for corresponding wind speed and maximum lift coefficient for a NACA0021 airfoil design

In order to receive the maximum lift coefficients for other wind speeds, the corresponding Reynolds numbers are calculated and inserted into Javafoil. The results can be seen in Fig. 3, right.

The tail section is designed as a NACA0018 airfoil section. This airfoil was chosen because of the good lift to drag and stall characteristics. Moreover, the airfoil section is not too sensitive to varying wind directions which makes it suitable as part of a wingsail arrangement. The tail wing has a chord length of 0.5 m and a height of 1.5 m. The distance between the trailing edge of the main wing and the leading edge of the tail wing is 0.5 m. This results in turning radius of 1.75 m for the wing arrangement in neutral position.

Self-trimming of the wingsail-tail-setup means in mechanical terms that the moment the main wing generates has to be balanced by the moment of the tail. With L_w and L_t being the lifting force of the wing and the tail, respectively, the momentum equilibrium can be expressed as

$$M_{\oplus} = L_w \times d_w + L_t \times d_t \equiv 0, \tag{4}$$

where d_w and d_l are the distances from the aerodynamic center of the wing and tail to the center of mass of the wing-tail-setup. Here we have suppressed the effects of the drag forces, since their contribution to the momentum is negligible for small angels of attack, see Fig. 3.

As mentioned above, the self-trimming capability is achieved best by aligning the center of mass of the wing arrangement at the aerodynamic center of the main wing. That means we have $d_w = 0$ and Eq. 4 reduces to

$$L_t \cdot d_t = \frac{1}{2}\rho \cdot d_t \cdot A_t \cdot C_{L,t} \cdot v^2 = 0. \tag{5}$$

Here we have made use of the fact that the lifting force of the tail is calculated according to Eq. 2 with the lift coefficient $C_{L,t}$ of the tail instead the wing lift coefficient.

Observing that all quantities in Eq. 5 besides the lift coefficient $C_{L,t}$ are non-zero (in fact positive), one arrives at the condition

$$C_{L,t} = 0. \tag{6}$$

The lift coefficient $C_{L,t}$ can be estimated by using lifting line theory combined with thin airfoil theory. One receives (see e.g. [8] for more details)

$$C_{L,t} = \frac{2\pi \cdot AR_t}{AR_t + 2}\left(\alpha + \delta_t - \frac{\varepsilon}{2}\right) \tag{7}$$

where AR_t is the aspect ratio of the tail, α is the angle of attack for the main wing, δ_{Tail} is the tail incidence and ε is the far-field downwash angle at the main wing. At the tail, the downwash can be assumed to be half of the far-field value, which is

$$\frac{\varepsilon}{2} = \frac{C_{L,w}}{\pi \cdot AR_w} = \frac{2\alpha}{AR_w + 2}. \tag{8}$$

Here we have assumed that the lift coefficient $C_{L,w}$ for the wing follows along the same lines as $C_{L,t}$ in Eq. 7.

Combining Eqs. 6–8 the result is

$$\frac{AR_w}{AR_w + 2}\alpha + \delta_t = 0, \tag{9}$$

where we again have made use of the fact that the aspect ratio A_t of the tail in Eq. 7 is non-zero and can be neglected. As will be explained in the next section, this final Eq. 9 was used to define the main wing incidence for a given tail incidence when run the CFD simulations stationary.

5 CFD Simulations

In order to understand the dynamics of the wingsail and the wing/tail combination better and to study its behaviour under different conditions, Computational Fluid Dynamics (CFD) Simulations were used. Our first idea was to simulate the free rotating system under different wind directions and strengths. However, up to date we did not succeed to receive realistic results when simulating flow driven motion. The simulations gave almost exclusively unrealistic movement patterns with oscillatory behaviour, and no stable solutions could be found, even when the wing and tail section passed the stability criteria established in the previous section. The reason is probably the limited mesh size which should have been dramatically larger for the volume inside the simulation boundary. However, this would have exceeded the capacity of our computers. Also the flow being compressible could have something to do with the faulty results. Even the low Reynolds number effects described in Sect. 3 may have played a part in the faulty simulations.

Therefore, we decided to run the simulations without any flow driven motion but rather use the calculated momentum equilibrium Eq. 9 in order to define the main wing incidence for a given tail incidence and run the simulations stationary. When simulating stationary situations, incompressible flow could be used without the issues experienced previously, and thus the results should be closer to reality than previously.

During the last days, simulations have been done for different scenarios of 0, 10, 15, 18, and 21° angles of attack and the results will undergo further analysis. Figure 4 shows the velocity of the airflow around the wing arrangement at 15, 18 and 21° angles of attack. As seen in the figure the flow gets more disrupt and turbulent behind the wing as the angle of attack increases leaving a clear image of the wake. The increased velocity of the airflow over the airfoil creates a pressure gradient which in turn generates thrust, but gradually the turbulent behaviour takes over generating an increasing risk for stalling.

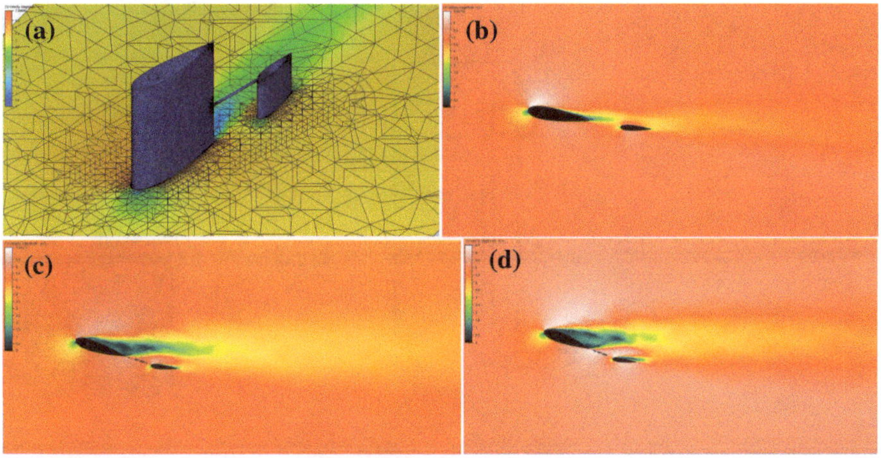

Fig. 4 Automatically generated mesh near the wing arrangement as well as in the far field. The main wing is positioned at 15° and the tail at 9.5° (**a**). The velocity of the airflow around the wing arrangement at an angle of 10° with a tail incidence of 6.3° (**b**). Main wing at an angle of 18° with a tail incidence of 11.5° (**c**). Main wing at an angle of 21° with a tail incidence of 13.5° (**d**). The initial wind speed for all four scenarios is 5 m/s

6 Roadmap for Future Work

So far, we have succeeded to simulate the airflow around a fixed position wing-tail system where we have calculated the respective angles of incidence. The next steps will be to study the results in more detail and to optimize the wing setup. Moreover, we are aiming to deepen our knowledge about CFD simulations and to gain access to a more powerful computer in order to receive true flow driven simulations. This will even open for the possibility to design an optimal airfoil section—which may not be a standard NACA profile—with superior lift-to-drag and stall characteristics. We are fully aware of the limits and weaknesses of CFD simulations and got, as we mentioned, in the beginning indeed non-plausible simulation results. However, our work is still ongoing, and an iterative approach leads us hopefully to more reliable results. The CFD results shall be used only in an indicative way in order to pave the way for the first prototype, which will finally decide about how reliable and successful our simulations are.

Concerning the more practical issues, our next step is to design the wingsail in detail and to build it. In this Section we will sketch some ideas and possible solutions to some of the challenges that will occur in the stage of planning and manufacturing the wing arrangement.

Aerodynamic and mass center alignment: The aerodynamic center on a symmetric airfoil is located precisely one quarter of the chord length from the leading edge of the airfoil. This is the point where the wing should be connected to the mast in order to ensure the self-trimming ability of the wing arrangement. The mass center however

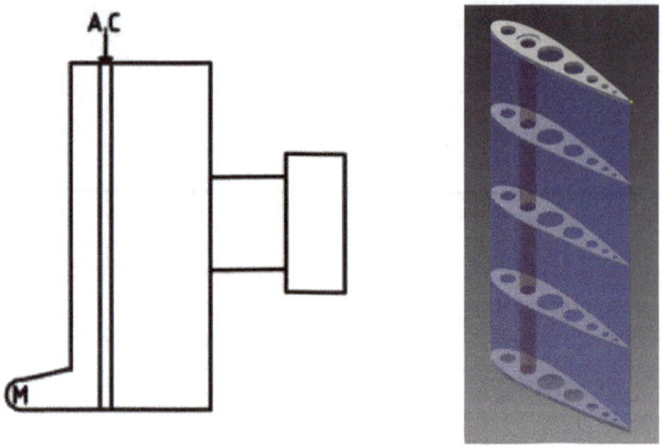

Fig. 5 *Left* Designed ballast method. *Right* A sketch depicting the construction of a wing

is located further back because a tailed wing configuration is tail heavy. This leads to the need of ballast in some form in front of the wing.

In principle, different ballasting methods could be used. The main issue with applying extra weight to a boat is the increased mass of the boat and—in the case of applying ballast high up the boat—the heightened center of mass of the boat itself. If the center of mass of the boat moves upwards too far, the boat will capsize or be unable to recover from big heeling forces. Having the mass center of the boat as low as possible is crucial to ensure the safety of operationality of any boat.

Figure 5, left, shows the optimal ballasting method for a wing sail. The "bulb" can be designed in a way that it doesn't contribute negatively to the aerodynamics of the sail. Another advantage achieved with the shown ballasting method is the fact that the counterweight is cased inside the wing (the bulb is a part of the wing). This opens up different uses for the ballast in the wing in the future. For example, the bulb could act as a housing for the electrical components and the batteries steering and power the tail actuator. Besides serving as ballast for the wing arrangement, this solution would also reduce the need for both batteries and ballast onboard the boat. The batteries can be loaded with solar panels fitted at the wing. If the electrical components steering the tail will have a wireless connection to the main computer inside the hull of the boat, no wiring what so ever would be needed to steer the wing.

Material and construction: As stated earlier the biggest disadvantage of the wingsail over conventional sails is the sails weight. Weight is a fundamentally important factor in ship design in order to ensure safety. Therefore, the wingsail will probably consist of a hollow construction as indicated in Fig. 5, right. In principle, many materials ranging from wood over carbon fiber and composites to aluminium/magnesium alloys and plastics could be used for ribs and the shell of the wingsail. The final decision has to take into account the durability of the light constructions of the wings. By choosing an optimal combination of materials and building method, a well functioning and weight optimized wing will be built.

References

1. Anderson, M.R.: Aerodynamic performance of a small-scale Wingsail Vessel. In: 54th AIAA Aerospace Sciences Meeting, AIAA SciTech, (AIAA 2016-2039) (2016)
2. Atkins, D.W.: The CFD assisted design and experimental testing of a wing-sail with high lift devices. University of Salford, Ph.D.-thesis (1996)
3. Cokelet, E.D., Meinig, C., Lawrence-Slavas, N., Stabeno, P.J., Mordy C.W., Tabisola H.M., Jenkins R., Cross J.N.: The use of Saildrones to examine spring conditions in the Bering sea. In: OCEANS 2015 - MTS/IEEE Washington, pp. 1–7. IEEE, Washington, DC (2015)
4. Cruz N.A., Alves, J.C.: Integration of wind propulsion in an electric ASV. In: Friebe, A., Haug, F. (eds.) Robotic Sailing 2015. Proceedings of the 8th International Robotic Sailing Conference, pp. 137–149. Springer, Heidelberg (2016)
5. Dahl, K., Bengsén A., Waller, M.: Power management strategies for an autonomous robotic sailboat. In: Fearghal, M., Dermot, T. (eds.) Robotic Sailing 2014. Proceedings of the 7th International Robotic Sailing Conference, pp. 47–55. Springer, Heidelberg (2015)
6. Domínguez-Brito, A.C., Valle-Fernández, B., Cabrera-Gámez, J., Ramos-de-Miguel, A., García, J.C.: A-TIRMA G2: an oceanic autonomous sailboat. In: Friebe, A., Haug, F. (eds.) Robotic Sailing 2015. Proceedings of the 8th International Robotic Sailing Conference, pp. 137–149. Springer, Heidelberg (2016)
7. Elkaim, G.H.: The atlantis project: a GPS-guided wing-sailed autonomous catamaran. Navigation **53**, 237–247 (2006)
8. Elkaim, G.H.: Autonomous surface vehicle free-rotating wingsail section design and configuration analysis. J. Aircr. **45**(6), 1835–1852 (2008)
9. Eriksson R., Fribe, A.: Challenges for autonomous sailing robots. In: 14th International Conference on Computer and IT Applications in the Maritime Industries, COMPIT15 Ulrichshusen, 11–13 May 2015
10. Hepperle, M.: About JavaFoil. Retrieved from Aerodynamics of Model Aircraft: http://www.mh-aerotools.de/airfoils/javafoil.htm. Accessed 8 May 2016
11. Melin, J: Modeling, control and state-estimation for an autonomous sailboat. Master thesis. Uppsala University (2015)
12. Melin, J., Dahl, K., Waller, M.: Modeling and control for an autonomous sailboat: a case study. In: Friebe, A., Haug, F. (eds.) Robotic Sailing 2015. Proceedings of the 8th International Robotic Sailing Conference, pp. 137–149. Springer, Heidelberg (2016)
13. Miller, P. H., Hamlet, M., Rossman, J.: Continuous improvements to USNA SailBots for inshore racing and offshore voyaging. In: Sauzé, C., Finnis, J. (eds.) Robotic Sailing 2012. Proceedings of the 5th International Robotic Sailing Conference, pp. 49–60. Springer, Heidelberg (2012)
14. National Aeronautics and Space Administration. Aerodynamic center. Retrieved from National Aeronautics and Space Administration: http://www.grc.nasa.gov/WWW/K-12/airplane/ac.html. Accessed 05 May 2016
15. Neal, M., Colin, S., Thomas, B., Alves, J.: Technologies for autonomous sailing: wings and wind sensors. In: Proceedings of the 2nd IRSC, Matsinhos, Portugal, 6–12th July 2009, pp. 23–30 (2009)
16. Rynne, P.F., Ellenrieder, K.D.: Development and preliminary experimental validation of a wind- and solar-powered autonomous surface vehicle. EEE J. Ocean. Eng. **35**(4), 971–983 (2010)
17. Sauzé, C., Neal, M.: MOOP: a miniature sailing robot platform. In: Schlaefer, A., Blaurock, O. (eds.) Robotic Sailing. Proceedings of the 4th International Robotic Sailing Conference, pp. 113–124. Springer, Heidelberg (2011)
18. Selig, M.S., Donovan, J.F., Fraser, D.B.: Airfoils at Low Speeds. H.A. Stockley, Virginia Beach (1989)

19. Shukla, P.C., Ghosh, K.: Revival of the modern wing sails for the propulsion of commercial ships. Int. J. Math. Comput. Phys. Electr. Comput. Eng. **3**(3) (2009)
20. Stelzer, R, Estarriola Dalmau, D.: A study on potential energy savings by the use of a balanced rig on a robotic sailing boat. In: Sauzé, C., Finnis, J. (eds.) Robotic Sailing 2012. Proceedings of the 5th International Robotic Sailing Conference, pp. 87–93. Springer, Heidelberg (2012)
21. Svenska 2.4mR Förbundet. BÅTEN. Retrieved from Svenska 2.4mR Förbundet: http://www.swe24metre.com/baten (2014)

Barlavento - Considerations About the Design of an Autonomous Sailboat

Pedro Castro Fernandes, Mario Monteiro Marques and Victor Lobo

Abstract Persistent monitoring of the ocean is important for several reasons, such as to: better understand the climate and ocean dynamics, improve navigation safety, make a better management of their resources and to prevent and combat abuse to the environment. However persistent monitoring can be extremely expensive. One way of decreasing costs is to use unmanned air, surface or underwater vehicles. However, energy autonomy is a major issue for this type of vehicles. For this reason, autonomous sailboats may be a good solution because they can collect renewable energy from the sea and atmosphere, thus being self-sustainable. The objective of this work is to develop and test an autonomous sailboat capable of performing persistent monitoring of the ocean. The Portuguese Naval Academy has been working on autonomous sailboats since 2010. However, the first autonomous sailboats, that used commercially available hulls, were not resistant enough to bad weather, had little available space for electronics, and were not very efficient. We now decided to design a radically different boat: a very thin and long monohull. We did so using freely available (or very low cost) software, low-cost off-the-shelf components, and simple 3D printers when necessary. This paper describes the creation of the hull using 3D CAD technologies and hydrodynamics simulation.

1 Introduction

There has been a lot of interest in developing unnamed systems persist monitoring of the oceans. The main applications of such systems are environmental monitoring, border and security control (mainly for preventing illegal immigration and smuggling), search and rescue, communication relay, etc. One of the main problems of

P.C. Fernandes (✉) · M.M. Marques · V. Lobo
Centro de Investigacão Naval, Base Naval de Lisboa, Almada, Portugal
e-mail: pedro.castro.fernandes@marinha.pt

M.M. Marques
e-mail: mario.monteiro.marques@marinha.pt

V. Lobo
e-mail: vlobo@novaims.unl.pt

J.C. Alves and N.A. Cruz (eds.), *Robotic Sailing 2016*,
DOI 10.1007/978-3-319-45453-5_2

such systems is energetic autonomy. For truly persistent ocean monitoring, the energy must be harvested from the environment. Many different approaches have been made, using solar power (e.g. Scout Boat [1]), wave power (e.g. waveglider [2]), or wind-power (e.g. FASt [3], AEOLUS [4], etc.). Windpower systems, and conventional sailboats in particular, have proven to be particularly fragile in ocean environments, particularly if they are small, and most have followed conventional designs used in yachts or larger boats.

The Portuguese Navy Research Center (CINAV), after some initial trials with conventional and commercially available model sailboats, started to develop a radically different design for an autonomous oceangoing sailboat. The aim is that sailboat be able to navigate independently of any human intervention. Sail is very dependent on the environment and is influenced by several factors, but the wind and the curl are the most important.

The wind speed and its direction defines the efficiency of the sails. The air flow passing through the boat sails generates two major forces: drag and lift. The conjugation of the boat's head, sail's angle, wind speed and direction, dictate the amount of drag and lift forces generated in each sail [5]. The shape, size, attack angle and number of sails also influence this result.

The curl affects the stability and speed of sailboat. By splitting the sea waves frequency spectrum in its components (see Fig. 1) x (forward - aft) and y (Starboard - Portside) [6], it is possible to take the following conclusions: (i) the components in y lead the ship to oscillate, and it may reach a roll angle beyond the maximum with positive stability; (ii) the components in x, if the direction of propagation of the wave is from forward to aft, leads to a decrease of speed, and otherwise leads to an increase of speed. The intensity with which those effects are felt depends not only on the sea, conditions, but also of the dimensions and type of hull [7].

Given the fact that environmental factors cannot be controlled, it is important to develop the sailboat characteristics in other to achieve its purpose even in heavy weather conditions.

This article is organized as follows: In Sect. 2 we define the most important factors taken into account in the hull design, and define the shape and dimensions of the hull; in Sect. 3 we describe the design and hydrodynamic simulations performed; in the Sects. 4, 5 and 6 we discuss aspects related to the design of the keel, rudder and centreboard respectively; Sect. 7 contains aspects related to peripheral systems; and finally in Sect. 8, we draw some conclusions about the results obtained.

2 Specifications Definition and Sizing

In the design process, the priority was the sailboat's capacity to resist to unpredictable weather conditions, keeping the dimensions as small as possible so that it can be easily transported. We defined the sailboat length overall (LOA) to 2000 mm, and started from there.

The next step was to define the displacement (Δ) of the sailboat. The greater the mass, the greater the inertia. This quantifies the difficulty of the sailboat to change its motion status. In certain situations, high inertia can be a benefit. On one hand, the greater the inertia, the lower is the loss of speed caused, for instance, by waves. On the other hand, the greater the inertia, the greater is the difficulty of the sailboat to gain speed or to make turns. Taking this in account, we decided that the maximum displacement of the sailboat would be 20 kg. This value is expressively low, taking into account the LOA. The hull type, materials and the stability have to be carefully chosen, in other to satisfy this parameters and still be able to perform the mission. With the LOA and the mass of the system defined, it is possible to start the shape design.

The buoyancy force is equal in module and has opposite direction to the weight of the fluid displaced [8]. We know that the weight of fluid displaced is the product of its density (ρ), by the volume displaced (∇) and by the acceleration of gravity [9]. Having the gravity acceleration in both sides of the equation, the Eq. (1) simplifies as shown:

$$\Delta = \rho \nabla \tag{1}$$

Using this equation, we can compute the volume of the sailboat underwater.

If the only forces acting on the sailboat were the buoyancy and the weight on the same vertical, the boat would be in balance and no movement would happen. In the real world, this is very unlikely. Many forces act on the sailboat, the main ones being related with wind and ocean waves. Depending on the localization of the application point of those forces, moments are created that tend to roll the sailboat, being necessary an opposing moment to balance it. The relations between the upsetting moment and the righting moment define the behaviour of the boat at sea [10]. The higher the righting moment produced when the boat is subject to a roll angle, the better, since, this way, the boat will be able to handle bigger upsetting moments. One way to manage this is through the design of the hull shape and size. The axles system used is represented in Fig. 1.

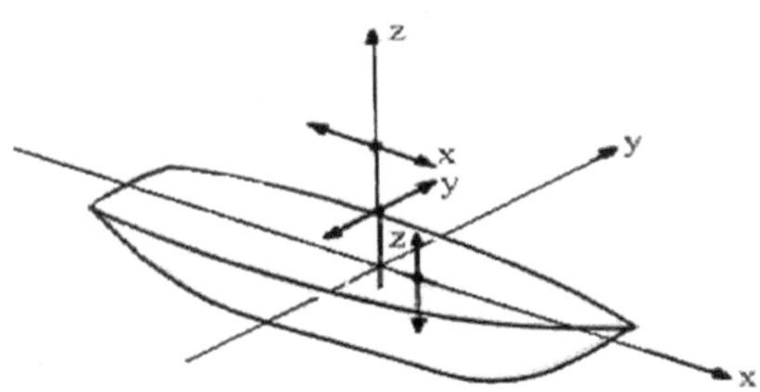

Fig. 1 Axles system

The goal is that small angles of roll produce large righting moments, meaning, the transverse metacentric height (Z_M) ought to be as high as possible. This can be expressed through the relation expressed in Eq. (2) [8].

$$Z_M = Z_B + \frac{I_y}{\nabla} \tag{2}$$

Where Z_B is fluctuation centre height and I_y is the second inertia moment of the fluctuation figure area.

For a given roll angle, the sailboat is in a stable condition if Z_M is higher than the gravity centre height (Z_G) [9]. This relationship is known as the stability condition and is represented in the Eq. 3.

$$Z_G < Z_M \tag{3}$$

Thus, the higher the Z_M and the lower the Z_G, the greater the stability. Z_M depends on the second inertia moment of the fluctuation area I_y, and can be calculated using the Eq. 4.

$$I_y = \int_A y^2 dA \tag{4}$$

This relation proves that the more area the sailboat has dispersed on the y axis, the higher will be the metacentric height, meaning it is beneficial to have a large beam. However, the length and the displacement of the sailboat are already set. If we want to maintain these properties and have a larger beam, we have to reduce on the draft. The draft is important in a sailboat. This type of boat relies on the wind to sail. The wind tends to drag the sailboat with it. A reaction force is required to propel the boat forward, instead of simply being dragged with the wind. This reaction force is generated between the water and the underwater body of the sailboat. The higher this force, the more it will maintain the desired course, resulting in a more precise sailing. In Fig. 2 we have a comparison between a bigger water drag force resistance and a smaller one. We consider that the force that propels the sailboat forward is the sum of the water drag resistance force and the wind force. We ignore other forces acting on it. In Fig. 2a we consider a larger drag resistance force then in Fig. 2b. It is evident that the angle between the forward force and the middle ship line is smaller in Fig. 2a, resulting in better sailing when compared with Fig. 2b.

Thus, we have to dimension the draft and the beam in order to maximize the stability and the lateral water drag resistance force.

The solution adopted was the hull in V shape [10], ensuring low hydrodynamic resistance and less speed loss caused by waves [10]. This form may not guarantee I_y high enough to have a balanced Z_M on the sailboat. In Eq. 3 we stated the stability condition. We can achieve positive stability either by higher the Z_M or either by lower the Z_G. As we're not able yet to calculate this value, we design assuming that with ballast management we can ensure a low value for Z_G, resulting in positive stability. To avoid the possibility of this not be enough and small upsetting moments developing big roll angles, we added two lateral hulls to the main one transforming the sailboat in a Trimaran.

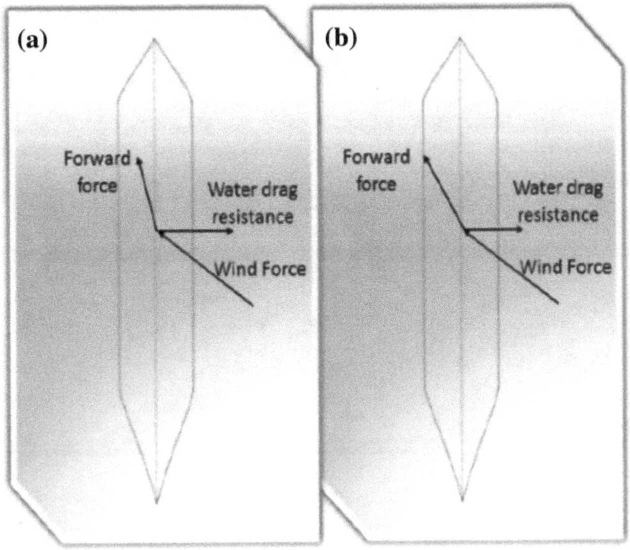

Fig. 2 Comparison between different water drag forces

If there is no interaction between the water flow around the different hulls of a trimaran, the total resistance is the sum of the resistance of each hull [11]. Keeping the resistance of each reduced hull, you get a small total resistance. This solution has the advantage of ensuring the sailboat's stability, keeping the hydrodynamic resistance low.

3 Design and Simulation in 3D CAD

We used the program DELFTship™ Free to design the main hull. This program is very versatile and user friendly, allowing to design the hull and calculate hydrostatic and hydrodynamics parameters easily.

The dimensions of the main hull are summarized in Table 1.

Figure 3 shows the main hull drawn on this software.

Making use of the same software, the hull drag with draft of 130 mm was calculated, which corresponds to the nominal situation of 12 kg of displacement. The program makes use of the resistance series of Delft [12] in this calculation. The graphic only contemplates the main hull drag, not counting with the drag caused by the rudder, hull roughness, keel and other underbody parts. This graphic is represented in Fig. 4.

We see that the forward drag is low, about 11 N at the speed of 4 Kts. The real drag is going to be higher than this. Nevertheless, the value still points to an efficient hull. For simplicity, we used the same hull shape for the lateral ones. This addition

Fig. 3 Main Hull designed in DELFTship™ Free

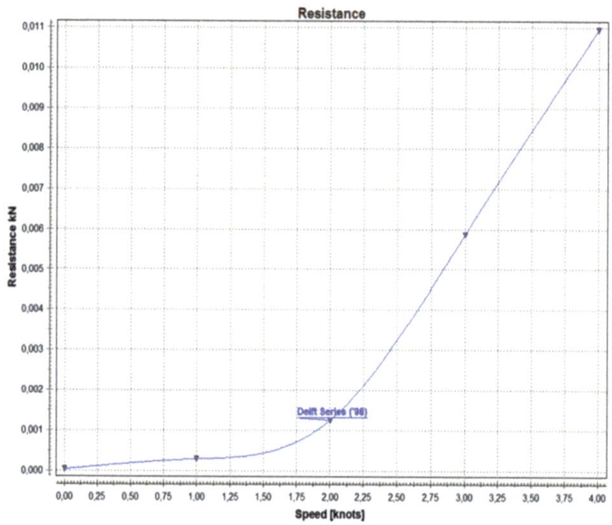

Fig. 4 Resistance as a function of speed for the main hull

Table 1 Hull main dimensions

Maximum beam	200 mm
Draft	130 mm
Length over all	1900 mm
Normal displacement	12 kg
Maximum displacement	20 kg
Cylinder block coefficient	0.1558
Prismatic coefficient	0.6272

increases the maximum displacement of sailboat over the 20 kg's fixed before. The function of the lateral hull is only to increase stability, so when the sailboat in upright the lower part of the lateral hulls is on the water line. The lateral hulls also provide a buoyancy reserve. When the sailboat gets a roll angle, the lateral hull on the heeled side submerge, producing a local buoyancy force. This force generates an upright moment. This can be seen as an increase in I_y, conducting to a bigger Z_M and better stability.

Having already defined all the dimensions and shape of the hull, we used the CAD program 3D SolidWorks Student Edition to draw all the parts of the sailboat. In this phase of the project, we stated that the mechanical resistance of the sailboat and the position of the centre of gravity are the major concern in order to achieve a robust and stable sailboat.

4 Keel Project

The first stage of the hull construction project was the keel design. The material used was rolled steel, since it has good mechanical resistance characteristics. However, this material has high density. Thus, we decided to use the minimum required to achieve a solid and robust part. On the other hand, the Z_G component contributes to the stability of the sailboat. Being $Z_G < Z_M$, the condition of stability, if the heavier materials are placed in lower positions of the sailboat, Z_G will also be lower, contributing to improve stability. It was defined that the keel would be constructed with "T" profile bars, with the dimensions shown in Fig. 5.

The "T" profile is widely used in structures for the relationship that it offers between resistances to loads verses quantity of material used. The aim is to transfer all loads applied to the sailboat to this part.

We defined that in the keel there would be three supports: two for the masts and one for the centreboard. Given major dynamic efforts in these components, these parts will also be constructed in rolled steel and directly welded to the keel.

The mast's supports are constituted by a vertical beam, welded in the bottom to the keel. On top of it, there is another beam, welded horizontally. This beam is welded to the mast guide. The mast guide is a tube with 25 mm external diameter, 1.5 mm thickness and 75 mm long. Concentrically to this tube, and directly welded to the keel, we have got another mast guide constructed with the same material and 50 mm long, leaving a gap of 75 mm between the tubes. This gap is going to be used

Fig. 5 Profile used in Keel

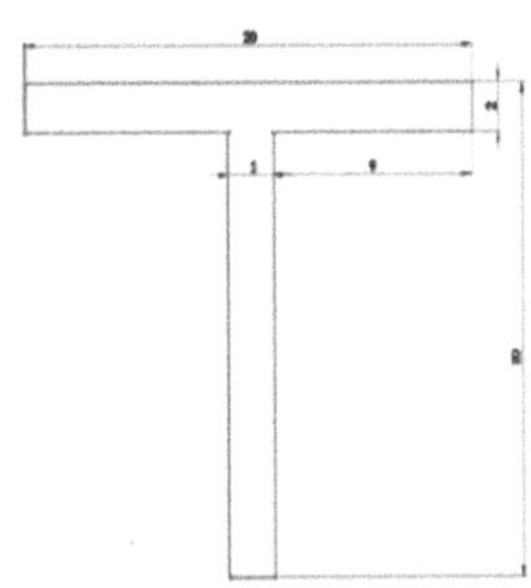

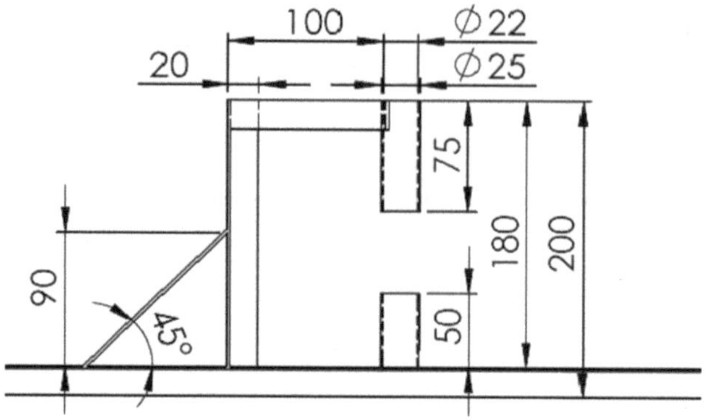

Fig. 6 Exploded view of mast support

to control the mast. Figure 6 is the front view of mast support. The keel contains also the support for the centreboard.

5 Rudder Project

The rudder plays an important role. It allows the sailboat to helm a course and increases the side drag, which is important so sail efficiently. The typical shape of the rudder is a foil.

By setting a certain angle to the rudder, the flow velocity on the two rudder faces will be different, generating a lift and a drag force. The lift force is responsible for generating a turning moment, and the drag force decreases the velocity of the sailboat. The decision of the foil shape was made using the program Designfoil Demo. This program can run simulations on a given foil at a certain angle, providing the lift and drag coefficient, and the pressure centre. The rudder profile is shown in Fig. 7.

The rudder size can be calculated by the Det Norske Veritas criterion [13], expressed in Eq. 5.

Fig. 7 Rudder Plant

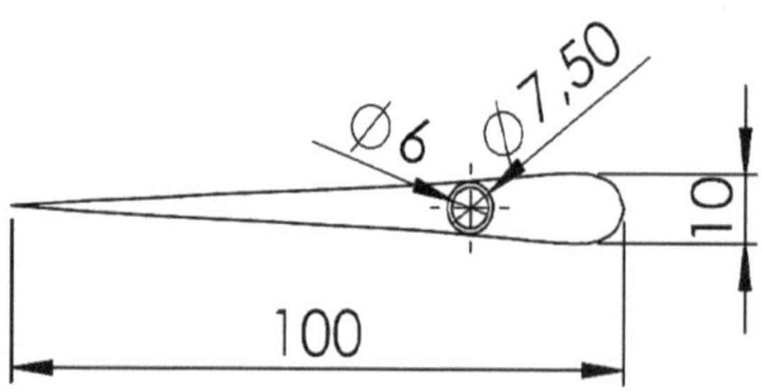

$$A_r \approx \frac{d.L_{pp}}{100}\left\{1+25\left(\frac{B}{L_{pp}}\right)^2\right\} \tag{5}$$

In which: A_r = rudders area; L_{pp} = Length between perpendiculars; B = beam and d = draft.

This equation provides the minimum rudders area. If the rudder is not placed directly behind the propeller, the area should be increased at least 30 %. The equation result for this sailboat is 48 cm^2. The rudder size should be well-adjusted because, on one hand, if it is too small it won't be able to turn the sailboat; on the other hand, if it is too big, it will generate too much drag force, slowing down the boat.

The rudder is in permanent contact with the water. This could be a problem because it's control would normally involve a hull passage under the waterline allowing water to go inside the sailboat. To minimize this risk, the guide of the rudder inside the sailboat goes higher than the water line, and has rubber retainers on top and bottom of the rudders guide.

6 Centreboard Project

It is intended that the centreboard has a high aspect ratio, because the higher this coefficient, the less drag is generated, maintaining the lift high enough to balance the lateral force generated by the sails [14]. Without this component, the sailboat would only be able to go in directions down wind. The efficiency of the centreboard is proportional to the aspect ratio. Also, for stability it is beneficial to have a big aspect ratio since need less ballast to achieve a lower Z_G position. The foil used in this component is shown in Fig. 8.

This component is 1000 mm long. It has to resist to the dynamical loads. In its construction, we decided to use the same material of the keel. This heavy material can resist to the loads, and it lowers the Z_G. This part measures 1000 mm and will have 3 kg of ballast on its edge.

In order to simplify the transportation of the sailboat, this component can be divided in two parts, one welded to the keel, 125 mm long, and other 875 mm long that can be attached to the first part by a screw joint. The keel is represented in Fig. 9.

The bulb will be located in the end of the centreboard and will be projectile type, as shown in Fig. 10.

Fig. 8 Centerboard profile.

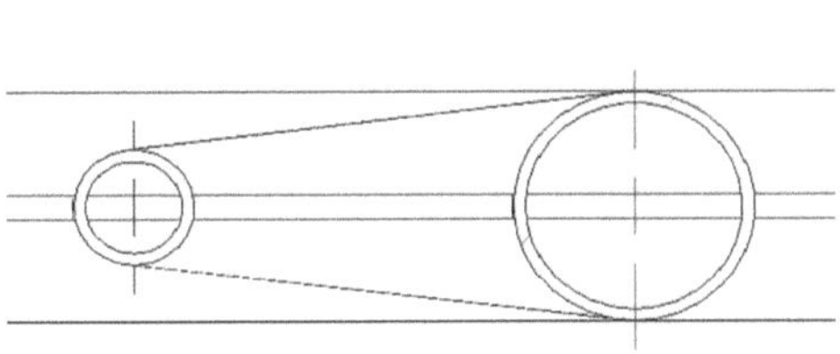

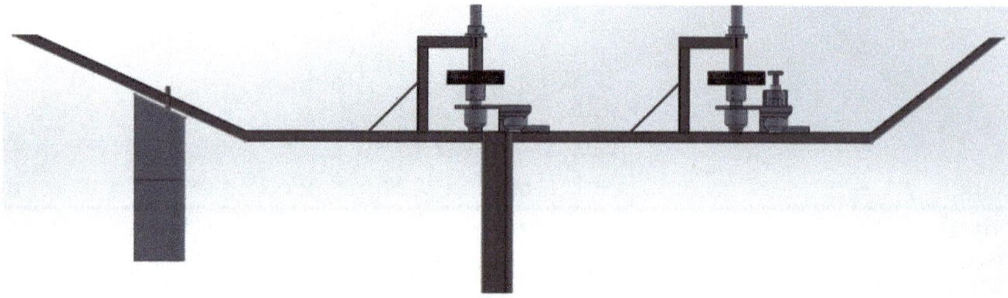

Fig. 9 Keel drawn in Solidworks Student edition

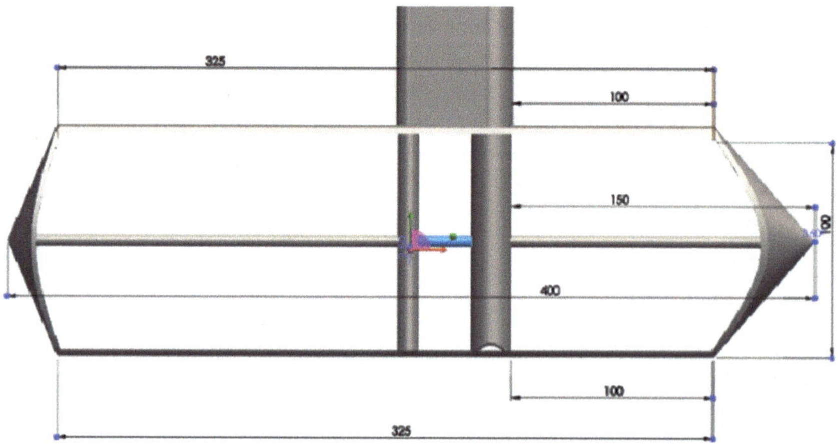

Fig. 10 Bulb

7 Peripherals Systems

We intended to create an improvement capable platform, and so developed a system that easily allows the exchange of components, such as mast and rudder, in case of material damage or for testing different components.

The control of the mast angles is made using a potentiometer. This potentiometer is linked to the mast by a gear. This system was developed in 3D CAD and printed in PLA. An electrical motor with a planetary gearbox was used. This system has a stall torque of 25 kg/cm. The torque is elevated by a set of gears also printed in PLA with a ratio of 2,14:1. The final result is shown in Fig. 11.

We use an Arduino Mega 2560 in the control centre. Attached to it we have got a compass, GPS, anemometer, 100 mw RF communications, and other sensors as shown in Fig. 12.

These sensors transmit the necessary information to the control algorithm. The waypoints coordinates can be saved directly to the SD Card, or can be sent by RF in a formatted message. The algorithm follows the waypoints sequentially, calculating

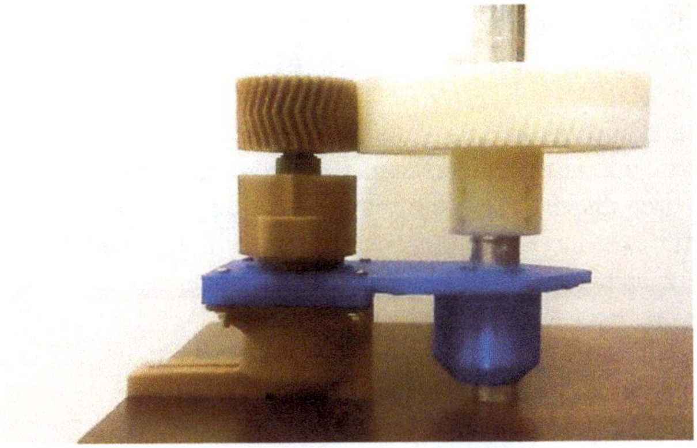

Fig. 11 Mast control System

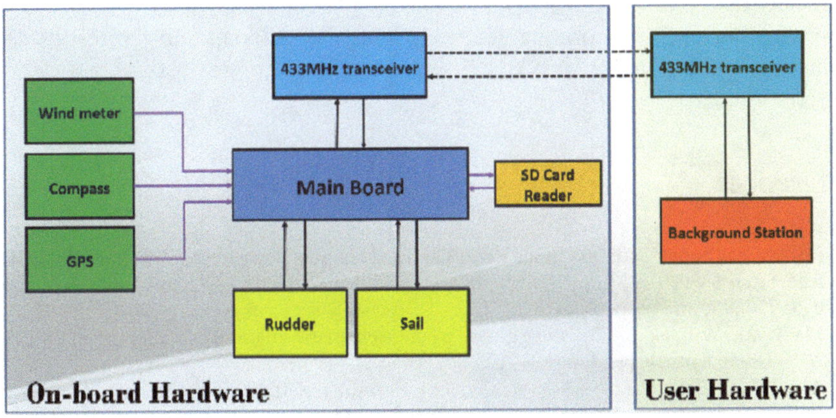

Fig. 12 System Architecture

the azimuth the sailboat should go by the GPS coordinates, adjusting the head by the compass readings, and the sails by the anemometer information.

8 Conclusion

This work is about the construction project of a new sailboat for CINAV. We recognized the importance of autonomous sailboats [3] and the importance to develop new approaches to this problem.

We started this article by talking about some concerns taken into account in the sizing the hull of the sailboat. We defined the LOA, the beam and draft of the sailboat. Then, we discussed the design and simulations done in 3D CAD, in order to achieve an efficient hull shape. After that, we explained the project of the keel, rudder and centreboard. We discussed the peripherals used and finally we end this article with some conclusions.

It is important to develop an efficient underbody shape in order to achieve low drag, but generate enough lateral resistance to avoid the sailboat simply drifting with the wind. The area dispersion in the y axis plays an important role in the sailboat stability. The aim is to have the most area dispersed in this direction, in order for the metacentre to be as high as possible. Another way to increase the sailboat stability is to lower the Z_G.

The draft and the centreboard are very important in sailing. Those components have to generate enough lift to balance the sails lateral force. The rudder not only provides steering to the sailboat, it also contributes to avoid the sailboat to drift with the wind. The rudder can by sized by the Det Norske Veritas criterion. This criterion gives a minimum value to the rudder size, in function of the boat size and characteristics.

We are currently performing sea trials with the sailboat, and the results seem promising.

References

1. Danko, P.: 'Scout,' Robotic Solar Boat, On Transatlantic Voyage Thanks To Group Of College Students (2013). http://www.huffingtonpost.com/2013/07/10/scout-robotic-solar-boat_n_3575669.html
2. Manley, J., Willcox, S.: The wave glider: a persistent platform for ocean science. In: OCEANS 2010 IEEE, Sydney, pp. 1–5 (2010)
3. Alves, J.C., Cruz, N.A.: FASt - An autonomous sailing platform for oceanographic missions. In: Proceedings of the MTS-IEEE Conference (2008)
4. Tranzatto, M., Liniger, A., Colombino, M., Hesse, H., Grammatico, S., Wirz, J.: AEOLUS, the ETH Autonomous Model Sailboat, Automatic Control Laboratory, Zurich, Switzerland (2014)
5. Wilson, R.M.: The Physics of Sailing. JILA and Depart-ment of Physics, Colorado. University of Colorado, Boulder (2010)
6. Gleason, B., Hallett, A., Osgood, D.H.: Multivariable Calculus. Wiley, New York (2005)
7. Roth, H.: Handling Storms at Sea. Adlard Coles, London (2009)
8. Lautrup, B.: Physics of continuous matter. The Niels Bohr Institute, Copenhagen (1998–2010)
9. Lewis, E.V.: Principles of Naval Architecture - Volume I. Stability and Strength. The Society of Naval Architects and Marine Engineers, Jersey (1988)
10. Misra, S.C.: Design Principles of Ships and Marine Structures. CRC Press, Taylor & Francis Group, Florida, United States (2015)
11. Mizine, I., Karafiath, G., Queutey, P., Visonneau, M.: Interference Phenomenon in Design Of Trimaran Ship, Greece (2009)
12. Keuning, J.A., Sonnenberg, U.B.: Developments in the Velocity Prediction based on the Delft Systematic Yacht Hull Series, Portsmouth, 1132-P (1998)
13. Journée, J.M.J., Pinkster, J.: Introduction in Ship Hydromechanics.: Delft University of Technology (2002)
14. Vacanti, D.: Keel and Rudder Design. Professional BoatBuilder, no. 95 (2005)

Part II
Mission Planning and Route Control

Autonomous Sailboat Local Route Planning

Mengqi Kang, Jinsong Xu, Jianyun Xu and Mingshu Du

Abstract The Velocity Made Good method was applied for local route planning of autonomous sailboat. A group of sailing tests on the lake verified the real-time performance and effectiveness of the Velocity Made Good method even when the wind field changes frequently.

1 Introduction

Marine environment monitoring and data measurement play an important role in weather forecast, disaster prevention, and ocean transportation. The use of autonomous sailboats for this purpose could achieve a better performance compared with the conventional methods such as oceanographic buoys, powered surface vehicles, aircrafts, and satellites. The major advantages include large range of monitoring, long term of period, high precision of data, low labor cost, and sustainable wind energy [12]. The successful operation of an autonomous sailboat relies on at least three modules, which are route planning, collision avoidance, and track following control [13].

United States Naval Academy generated the north and south route for the sailboat Spirit of Annapolis in North Atlantic Ocean according to the pilot charts of NOAA [7]. Ulm University utilized Constrained Handling Rules and A* algorithm to develop a long-term route planning for ASV Roboat [8]. University of Auckland modeled the uncertainty of the weather by a branching scenario tree and finally found a route with

M. Kang (✉) · J. Xu · J. Xu · M. Du
State Key Laboratory of Ocean Engineering,
Shanghai Jiao Tong University, Shanghai, China
e-mail: gertie.kang@sjtu.edu.cn

J. Xu
e-mail: jingsong@sjtu.edu.cn

J. Xu
e-mail: xujianyun101@sina.com

M. Du
e-mail: dms1415@163.com

J.C. Alves and N.A. Cruz (eds.), *Robotic Sailing 2016*,
DOI 10.1007/978-3-319-45453-5_3

the minimized sailing time [10]. A research team in Spain developed a route planning program based on Nelder–Mead Simplex algorithm [1]. Hendrik Erckens developed a navigation system for an autonomous sailing vessel Avalon, which employed A* algorithm to generate the fastest path to a given destination [3]. Ocean University of China achieved the dynamic route planning based on Bellman optimality principle and fuzzy concept [5, 6].

In this research, the Velocity Made Good method was applied to realize the local route planning for an autonomous sailboat. The model sailing tests on the campus lake verified the real-time performance and effectiveness of the Velocity Made Good method even when the wind field changes frequently.

2 Principles of Autonomous Sailboat Intellectualization

Figure 1 illustrates the program frame of the autonomous sailboat. Firstly, the boat speed performance can be predicted from velocity prediction program (VPP) with the wind forecast results or measured wind data. Secondly, the route planning program should determine a safe and fast route according to the speed prediction and sailboat position. Finally, the instructions of rudder angle and sail angle could be generated from the track following program.

For autonomous sailboats, VPP result is the foundation for route planning and track following control. The VPP program can predict the maximum boat speed in various headings and wind speed based on dynamic equations involving the aerodynamic model of the sail and the hydrodynamic model of the hull. The sail aerodynamic model can be obtained from wind tunnel experiments, CFD computations, or empirical formulae. The hull hydrodynamic model can be obtained from towing tank experiments, CFD computations, or empirical formulae [9].

The complete route planning of the sailboat consists of global route planning, long-term planning, and local route planning, which corresponds to several-week voyage, several-day voyage, and three-hour voyage respectively [7]. The intermediate points of global routing are the goal points of long-term planning, and the intermediate

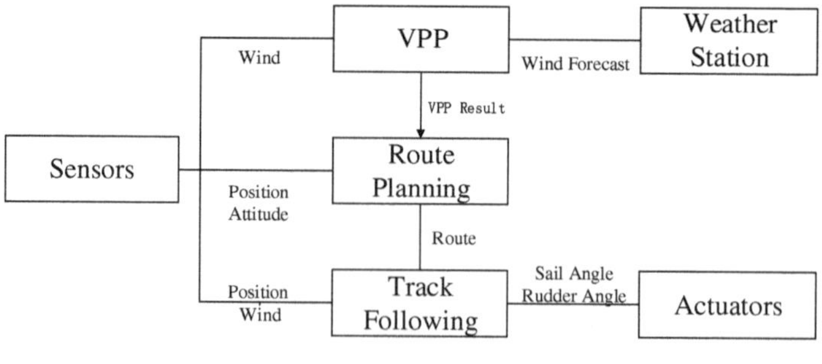

Fig. 1 Program frame of the autonomous sailboat

points of long-term planning are the goal point of local route planning. Global routing and long-term planning must rely on the meteorological record and weather forecast. Long-term planning should be renewed periodically according to the weather forecast updates every three hour. However, local route planning should be updated in real time with the change of the measured wind data.

Track following control program should guide a sailboat to sail along a planned path by controlling sail angle and rudder angle continuously. E.C. Yeh developed a controller based on fuzzy logic theory to ensure the maximum speed on specific path [16]. Roncin Kostia and Jean-Michel Kobus used PID theory to simulate two sailing boats in match racing [11]. N.A. Cruz utilized P/PI controller to tack when the sailboat sails against the wind [2]. Qian Wang developed track following controller with the combination of a sail controller based on general rules and a rudder controller based on fuzzy logic theory [15].

3 Autonomous Sailing Tests

3.1 Sailboat Model and Hardware Integration

The main dimensions of the sailboat model is shown in Table 1 and Fig. 2, and more design details can be referred to Ref. [15]. As shown in Fig. 3, the sensors equipped in the sailboat include GPS, AHRS, anemoscope, and sail angle encoder, etc. The actuators include rudder steering engines and sail winches. There are other equipments such as remote control devices and WiFi routers. Intercommunication between computer on the bank and microcomputer onboard was achieved by network. The measured data from the sensors were transferred to the computer on the bank. The sail angle and rudder angle instructions were generated from route planning and track following program in computer on the bank, and transferred to the microcomputer onboard model.

Table 1 Main particulars of the hull

Overall Length (m)	1.500
Waterline Length (m)	1.311
Beam (m)	0.476
Waterline Breadth (m)	0.364
Molded Depth (m)	0.433
Displacement Volume (m^3)	0.015
Draught (m)	0.069
Wetted Surface (m^2)	0.493
C_P	0.563

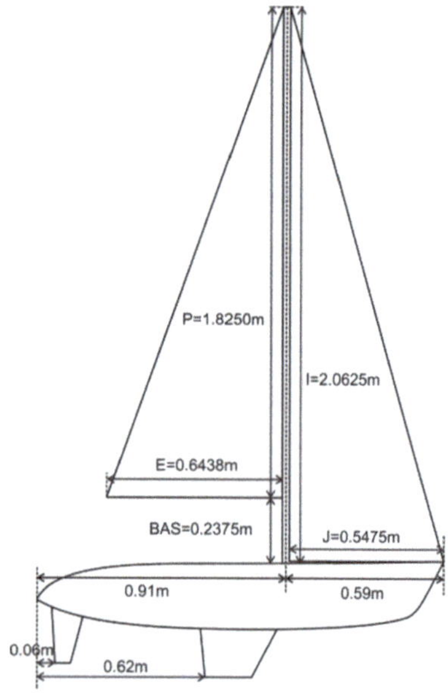

Fig. 2 Dimensions of the sailboat model [15]

Fig. 3 Sailing tests scenery

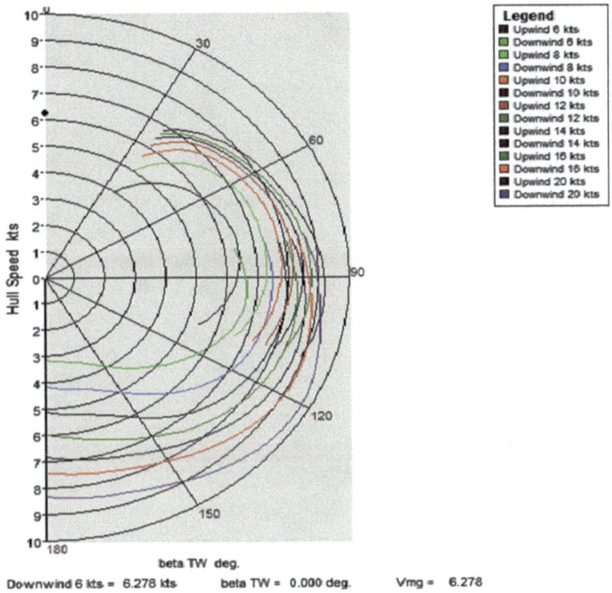

Fig. 4 Polar diagram of the VPP results

3.2 *Speed Polar Diagram*

Figure 4 shows the velocity prediction results obtained from module SPAN of MAXSURF program. In the diagram, each curve shows the maximum speeds the sailboat could reach in various course directions with the specified wind speed. The maximum speed along any course direction in any wind condition can be obtained by bilinear interpolation of the speed curves.

3.3 *Framework of Controller*

As shown in Fig. 5, the whole controller was divided into four modules, which are wind direction module, judgment module, local routing module, and track following controller. The function of the wind direction module is to filter out the influence of hull rolling on wind direction measurement. The judgment module compares the current position of the sailboat with the goal point. The local route planning module will determine the next goal point when the sailboat reaches the current goal point. The track following controller can generate the instructions of the sail angle and rudder angle for path control. Figure 6 shows the VeriStand interface of our sailing tests.

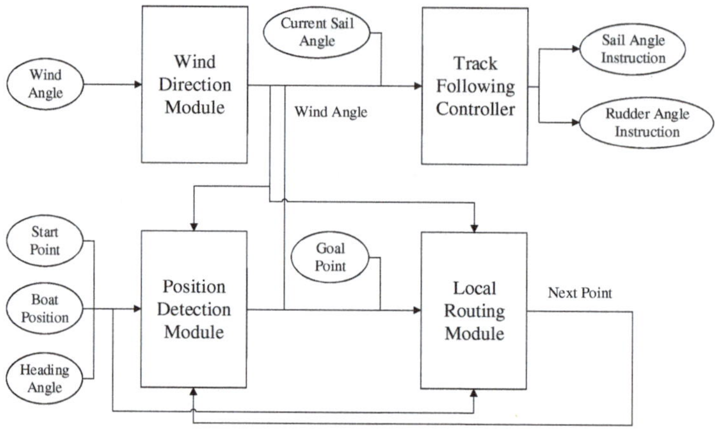

Fig. 5 Framework of controller

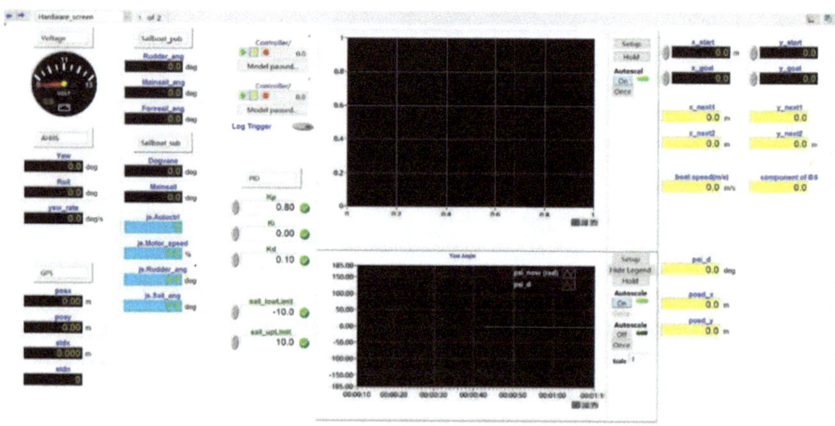

Fig. 6 VeriStand interface of the sailing tests

3.4 *Local Routing Module*

Since the wind field changes frequently on the lake, the Velocity Made Good method [14] was applied in this research. The core element of the method is to determine the course direction at any current point. The best course direction should maximize the velocity component towards the goal point to achieve the shortest voyage time. The method can update the course direction in real time according to current wind data and current position of the sailboat. The route obtained from the Velocity Made Good method is not the global optimization result. However, it can satisfy the requirements of sailboat local routing in any wind conditions.

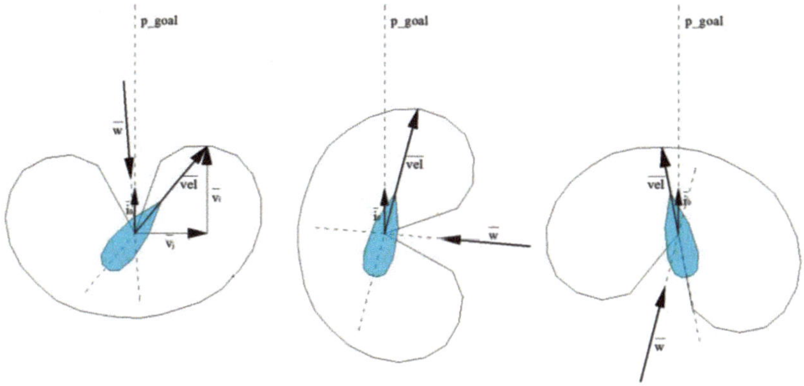

Fig. 7 Illustration of the Velocity Made Good method [14]

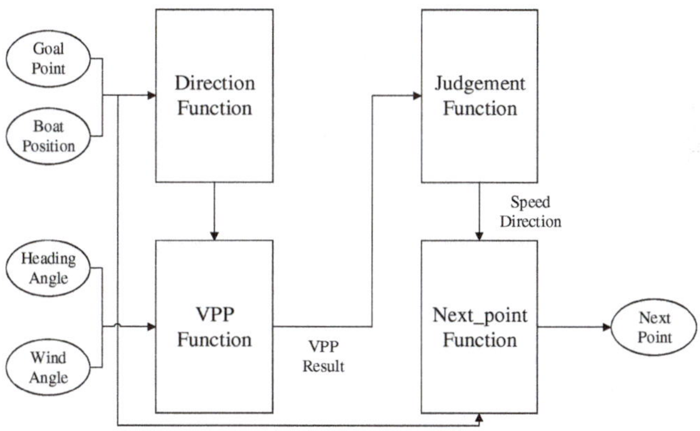

Fig. 8 Local routing module

As shown in Fig. 7, $\overrightarrow{vel}$ is the achievable boat velocity for any specified course direction, $\overrightarrow{\iota_0}$ is the vector aiming at the goal point, $\overrightarrow{w}$ is the vector of true wind. The value of $|\overrightarrow{vel}|$ can be obtained from VPP given the wind speed $|\overrightarrow{vel}|$ and the angle $|ra - ha|$ between wind direction and course direction. Thus the velocity component $\overrightarrow{v_l}$ towards the goal point can be obtained as follows,

$$|\overrightarrow{vel}| = vpp(wv, |ha - ra|)$$

$$vi = \overrightarrow{vel} \cdot \overrightarrow{\iota_0}$$

As shown in Fig. 8, the input variables include coordinate values of the goal point, current point of the sailboat, attitude information, and apparent wind direction. The output variables include next two goal points and speed prediction.

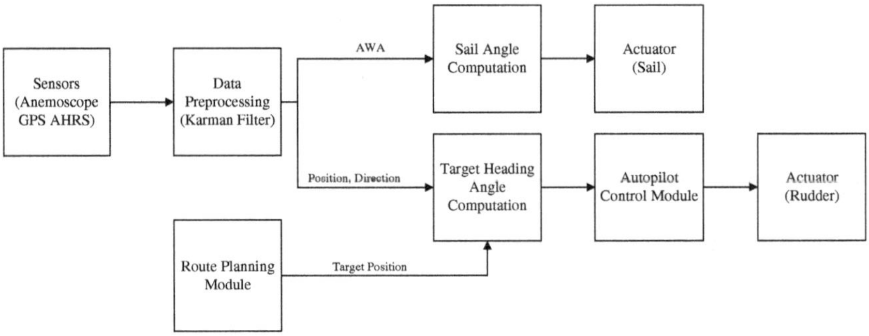

Fig. 9 Track following controller

3.5 Track Following Controller

The track following controller adopted the PID theory and line of sight method to build the autopilot module and tracking control module. In line of sight method, the control objectives were several discrete points between the start and goal points instead of a definite route, and the heading angle of the sailboat was the only control parameter [4].

As shown in Fig. 9, the autopilot control module was based on PID theory. The rudder angle was set according to the difference between the target heading angle and current heading angle, while the sail angle was set according to the apparent wind angle and difference of heading angle.

4 Sailing Test Results

The model sailing tests were conducted on the campus lake of Shanghai Jiao Tong University. The wind forecast that day was northeast wind at about 10 knots. However, the true wind field changed frequently due to the buildings around the lake.

Figure 10 shows the downwind and upwind sailing results from both sailing tests and simulation. The paths of sailing tests are shown in bold, and the simulated results according to actual wind data are shown in thin lines. The simulation paths are consistent with the sailing test results. There are differences between voyage time of the simulation and sailing tests, because the wind speed was assumed to be constant 10 knots in the simulation.

Figures 11 and 12 plot the time history curves of the boat position, boat speed, yaw rate, heading angle, rudder angle, and sail angle when sailing downwind and against the wind separately.

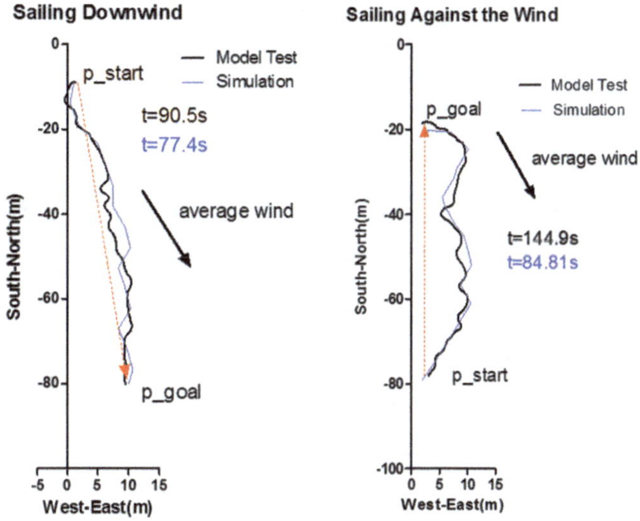

Fig. 10 Comparison of sailing tests and simulation results

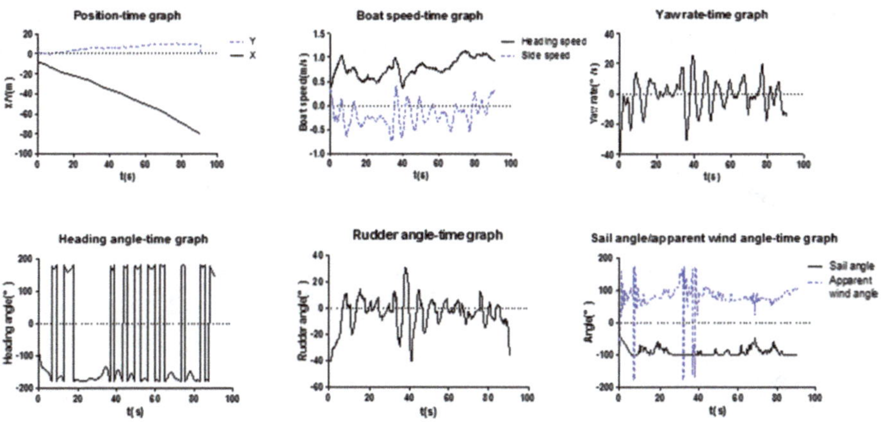

Fig. 11 Test results of the sailing downwind

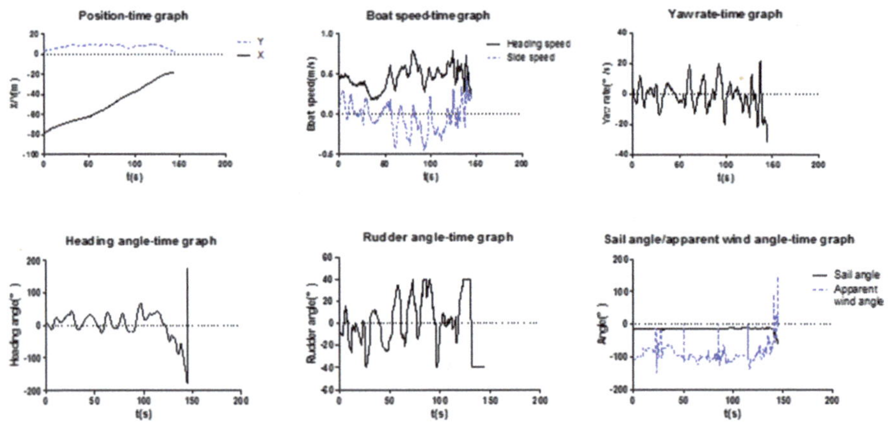

Fig. 12 Test results of the sailing against the wind

5 Conclusions

The model sailing tests on the campus lake verified the real-time performance and effectiveness of the Velocity Made Good method. The method can work well with the track following controller, and meet the requirements of local route planning even when the wind field changes frequently.

References

1. Cabrera-Gámez, J., Isern-González, J., Hernández-Sosa, D.: Optimization-based weather routing for sailboats, Robotic Sailing, et al.: Springer. Berlin **2013**, 23–33 (2012)
2. Cruz, N.A., Alves, J.C.: Auto-heading controller for an autonomous sailboat. In: OCEANS 2010 IEEE - Sydney, pp. 1-6 (2010)
3. Erckens, H., Busser, G.A., Pradalier, C., et al.: Avalon: navigation strategy and trajectory following controller for an autonomous sailing vessel. Robot. Autom. Mag. IEEE **17**(1), 45–54 (2010)
4. Fossen, T.I.: Handbook of Marine Craft Hydrodynamics and Motion Control [M], pp. 242–243. Wiley (2011)
5. Ge, Y.: Research on Path Planning for Sailboat Racing Based on Fuzzy Logic and Evolutionary Algorithms. Oecan University of China, Shandong (2005)
6. Ge, Y., Meng, Q., Wei, Z.: Optimum path planning method for straightway sailing race based on dynamic programming. Control Decis. **20**(12), 1360–1364 (2005)
7. Gibbons-Neff P., Miller P.: Route Planning for a Micro-transat Voyage. In: 4th IRCR, pp. 183-194 (2011)
8. Langbein, J., Stelzer, R., Frühwirth, T.: A Rule-based Approach to Long-term Routing for Autonomous Sailboats//robotic Sailing. Springer, Berlin (2011)
9. Larsson, L.: Scientific method in yacht design. Annu. Rev. Fluid Mech. **22**, 349–385 (1990)
10. Philpott, A., Mason, A.: Optimising yacht routes under uncertainty. In: The 15th Cheasapeake Sailing Yacht Symposium (2001)

11. Roncin, Kostia, Kobus, Jean-Michel: Dynamic simulation of two sailing boats in match racing. Sports Eng. **7**(3), 139–152 (2004)
12. Rynne, P.F., von Ellenrieder, K.D.: Unmanned autonomous sailing: current status and future role in sustained ocean observations. Mar. Technol. Soc. J. **43**(1), 21–30 (2009)
13. Stelzer, R.: Autonomous Sailboat Navigation. De Montfort University, The United Kingdom (2012)
14. Stelzer, R., Tobias, P.: Autonomous sailboat navigation for short course racing. Robot. Auton. Syst. **56**(7), 604–614 (2008)
15. Wang, Q., Kang, M., Xu, J., Xu, J.: Autonomous sailboat track following control. In: Proceedings of the 8th International Robotic Sailing Conference, pp. 125–136 (2015)
16. Yeh, E.C., Bin, J.C., Fuzzy control for self-steering of a sailboat. In: Proceedings of Singapore International Conference on of Intelligent Control and Instrumentation, SICICI & #039; 92. IEEE, **1992**, 1339–1344 (1992)

Improving Instrumentation Support and Control Strategies for Autonomous Sailboats in a Regatta Contest

Luís Gomes, Anikó Costa, David Fernandes, Hugo Marques and Filipe Anjos

Abstract This paper presents a controller architecture targeted for autonomous sailing of a small yacht, having in mind its control within a regatta contest. The controller architecture considers a three layer hierarchical decomposition, where the bottom layer takes care of the low level control of the sail and the rudder, based on the usage of a fuzzy controller, while the middle one selects the adequate navigation strategy (avoiding no-go zone), and the top layer is responsible for the definition of the regatta sequence of intermediate and final goals. A model-based development strategy is used, relying as much as possible on automatic code generation from models. Initial electronic instrumentation support comes from a compass, a GPS and a wind vane, but current developments include addition of an anemometer (providing wind speed and direction), as well as the introduction of virtual sensors obtained from computation of sensor data (as velocity obtained from GPS). A framework allowing the emulation of the different electronic instrumentation components support a complete offline validation of the controller, in a laboratory environment.

1 Introduction

The present paper addresses the development framework for a sailboat controller, starting from a sailboat built for RC control, where the RC control part was replaced by a controller board and adequate electronic instrumentation. The selected sailboat is

L. Gomes (✉) · A. Costa
FCT and UNINOVA, CTS, Universidade Nova de Lisboa, Lisbon, Portugal
e-mail: lugo@fct.unl.pt

A. Costa
e-mail: akc@fct.unl.pt

D. Fernandes · H. Marques · F. Anjos
FCT, DEE, Universidade Nova de Lisboa, Lisbon, Portugal
e-mail: ds.fernandes@campus.fct.unl.pt

H. Marques
e-mail: hm.marques@campus.fct.unl.pt

F. Anjos
e-mail: f.anjos@campus.fct.unl.pt

J.C. Alves and N.A. Cruz (eds.), *Robotic Sailing 2016*,
DOI 10.1007/978-3-319-45453-5_4

a 1 m replica of America's Cup racing yacht, while the selected controller board is an Arduino Mega [1], and basic electronic instrumentation includes a compass, a GPS, and a wind vane. All together it is a low-cost sailboat amenable to support autonomous navigation (however without any collision detection or avoidance mechanism).

This work benefits from lesson learnt from previous works on autonomous sailboats as a result of several meetings and interactions with colleagues from other institutions, in particular from the FAST initiative from University of Porto - Faculty of Engineering [2, 3], as well as from several experiments with LaserRC yachts owned by Portuguese Navy School.

One of the goals of the work is to build a simple platform amenable to be used in attracting students to STEM (science, technology, engineering and mathematics) areas through specific initiatives, namely short summer courses for high school students.

The paper starts to present the developed controller, giving information on both the hardware and software architectures. Next section briefly presents computational tools that were used to develop the controller, while following section addresses new electronic instrumentation components under development, as well as an emulation framework to support laboratory full testing of the controller behavior. Finally, some conclusions and future works are presented.

2 Hardware and Software Architectures

This section is divided into two main subsections, while the first one briefly presents the hardware architecture, the second addresses the three layer hierarchical software controller. The descriptions expand and update previous work presented elsewhere [4].

2.1 Hardware Architecture

Selection of the hardware platform took into consideration several criteria, including overall low-cost, moderate power consumption, as well as easy of programming, even for non-experienced programmers (as high school students are also targeted users of the platform). There are several platforms satisfying those requirements, but the selection goes for an open hardware platform, which also has a strong community behind: the Arduino platform [1]. It is also important to note that this selection does not compromise usage of the proposed solution in other platforms (or by more experienced users), due to its modular nature and easy porting of the software solution.

A minimum set of electronic instrumentation components was connected to the Arduino Mega, including:

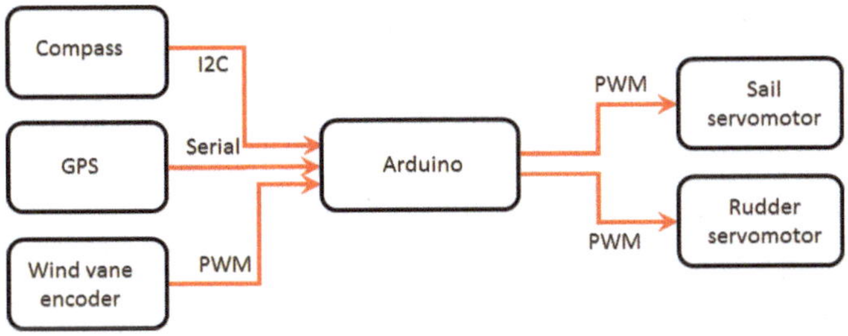

Fig. 1 Hardware subsystems interconnections and selected interface protocols

- a tilt compensated compass (providing also information from accelerometers, together with the pitch and roll from 3-axis gyro; I2C interface was elected for interconnection; the CMPS10 and CMPS11 modules have been used),
- a GPS receiver (the MediaTek MT3329 GPS V2.0 module was selected; serial communication was used), and
- the wind vane was homemade based on a contactless magnetic position encoder (the ams AS5040 Adapter board was selected, using PWM interface).

Two PWM controlled servomotors commonly used in RC control were used to control the sail and the rudder. A block diagram of the hardware architecture is illustrated in Fig. 1; it is similar to common architectures found in the literature, namely in [5].

The diversity of interconnection solutions allows a good coverage on common interfaces to interconnect hardware modules (and anticipate the use of the platform as a didactic set-up supporting embedded systems development teaching).

2.2 Three Layer Hierarchical Controller

As already referred, the goal is to support autonomous sailing within a regatta path contest. For that end, a three layer hierarchically structured controller is proposed, as presented in Fig. 2.

The top level is responsible for the mission definition, where a list of intermediate (considering the different buoys) and final goals is defined, establishing the list of segments to follow from the start line to the end goal finish line. Starting with the first segment in the list, after detection of goal accomplishment on that segment, the control will iterate on the segment list till reaching final destination.

The middle layer is responsible to elect the right navigation strategy for each segment of the path, taking into consideration current direction of the wind, current orientation of the sailboat, and intended direction of sailboat movement.

Finally, the lower layer is responsible for actuate sail and rudder. Following two subsections give further details on the two lower layers.

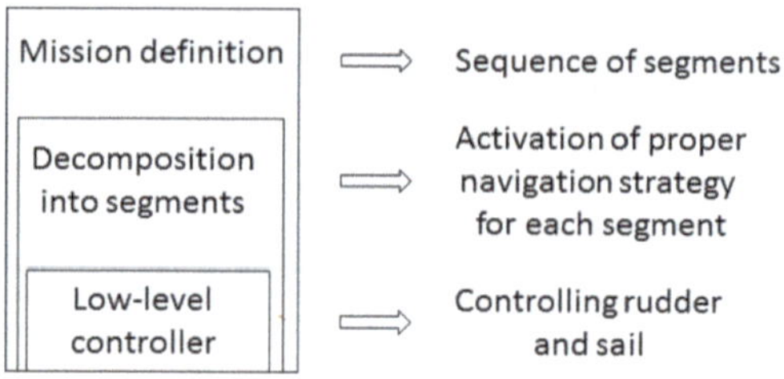

Fig. 2 Three layer structure of the controller

2.2.1 Defining Navigation Strategy

Each navigation segment is characterized by a set of points, namely initial point A, destination point B, as well goal line B1-B2. It is important to note that distance between B1 and B, and B and B2 (denominated as d2 and d3 respectively in Fig. 3) can be different. This characterization allows easy adaptation to different requirements, namely when the goal line B1-B2 represents the final goal line (and where d2 and d3 would be most probably equal), as well as intermediate goals (as the one going around a buoy, where the destination point B is not far from the buoy and B1 can impose a minimum distance to the buoy, and where B2 can allow some flexibility to adjust to specific navigation constraints).

Current navigation strategy defines an area where navigation is possible (regions 1 and 2 of Fig. 3), including close reach navigation. Starting with the definition of the desired optimal trajectory (line l0 in Fig. 3), two equally spaced parallel lines are defined considering distance d1 to l0 (lines l1 and l2 in Fig. 3). Our current setup considers d1 equal to 5 m (however, future works will study different values

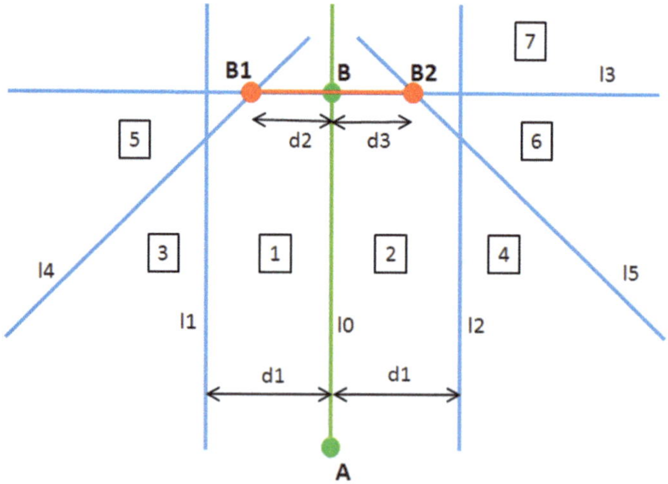

Fig. 3 Geometric characterization of navigation corridor considering initial position A, and goal line B1-B2

considering specific navigation situations). When approaching goal line, it is necessary to consider navigability minimum conditions (namely for close reach navigation), and two lines (14 and 15 in Fig. 3) were defined having a specific angle in relation with line 13 (which includes the goal line), and allowing detection of moving outside the navigation area. As a very conservative navigation strategy has been adopted, an angle of 45 degrees has been considered (however, future works will address improvements on this aspect, considering specific type of sailboat and speed).

In this sense, considering the geometry presented in Fig. 3, several regions are considered:

- based on splitting the area using goal line (line 13), generating region 7 and the others,
- based on splitting the area using vertical lines, namely regions 3, 1, 2, and 4, separated by lines 11, 10, and 12, respectively,
- based on splitting the area using lines 14 and 15 adequate to split between no-go zone and navigation area (used when the sailboat is in close reach navigation approaching the goal line).

Based on the presented geometric characterization and on wind direction related with goal direction, the adequate navigation mode is selected in order to avoid the no-go zone. Figure 4 summarizes the behavior using a state diagram notation.

When entering into any of the three referred states new initial point and destination point (final goal) for navigation direction are set according with the following:

- for `Regular navigation` state, initial and final point will be A and B, as initially defined for the segment;
- for state `Into the wind (port)` as well as for state `Into the wind (starboard)`, the initial point of the navigation direction is set to the current position of the sailboat, while the destination point is set to a point belonging to a line resulting from a sailboat tack (outside the navigation area), using a series of close-hauled segments to beat a course upwind.

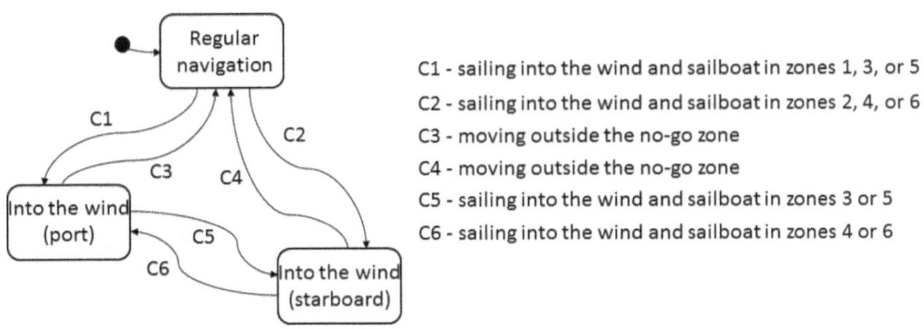

Fig. 4 State diagram governing navigation mode selection

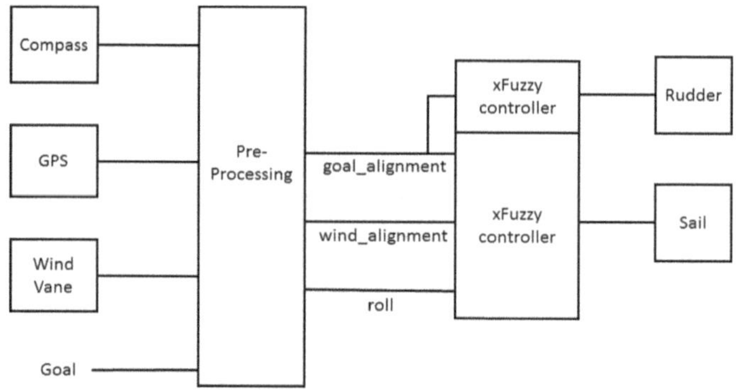

Fig. 5 Block diagram of the low level controller

2.3 *Low Level Fuzzy Controller*

Fuzzy control techniques were selected for the implementation of the low-level controller both for rudder and for sail, as in [6]. Main motivation is associated with the fact that it would be easier to describe behavioral dependencies using a set of rules (as reliable equation models are not available), as well as with this strategy it is possible to include non-experienced users to exercise different control strategies. A block diagram for the low-level controller is presented in Fig. 5.

A preprocessing block is responsible to obtain data from the compass (sailboat current orientation, as well as roll attitude), GPS (current location), and wind vane (direction of the wind in relation with sailboat orientation). This information was considered to provide most relevant dependencies and support development of a simple, although reliable, controller. The variables selected to be inputs of the fuzzy controllers are only three:

- `goal_alignment` (obtained by the difference between the sailboat orientation and the desired direction towards the goal); this variable is used by both rudder and sail controllers;
- `wind_alignment` (obtained by the difference between the wind direction and the sailboat orientation); this variable is used only by the sail controller;
- `roll` (obtained from compass module, and providing an indirect information on wind velocity); this variable is used only by the sail controller.

For the rudder controller, the `goal_alignment` variable is the only input, characterized as illustrated in Fig. 6, a set of fuzzy rules is defined as presented in Fig. 7, where output variable is characterized using five possible singleton values.

For the sail controller, a similar approach was followed, considering three input variables, namely `goal_alignment` (holding the seven fuzzy values presented in Fig. 6), `wind_alignment` (holding three fuzzy values: stern, slant, and bow), and `roll` (holding two fuzzy values: right and heel), while output fuzzy variable holds five possible singleton values (`full_loose`, `loose`, `half`, `tight`,

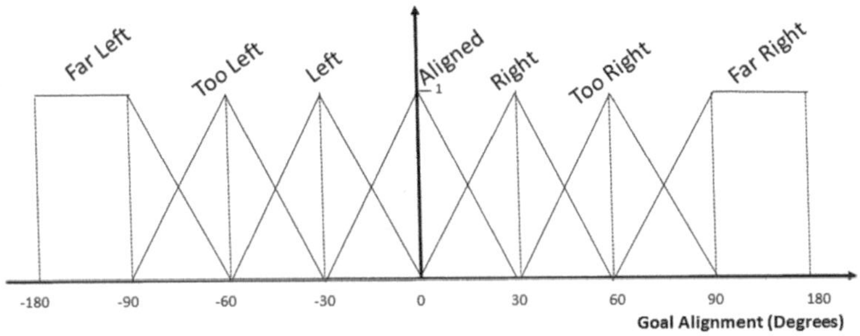

Fig. 6 Fuzzy sets associated with `goal_alignment` input variable

Fig. 7 Fuzzy rules associated with rudder controller

IF goal_alignment IS	THEN rudder IS
far_left	full_port
too_left	port
left	centered
aligned	centered
right	centered
too_right	starboard
far_right	full_starboard

`full_tight`). The controller is composed by a set of 42 fuzzy rules. A description of a former version of the controller (although satisfying the same structure) can be found in [4].

3 Development Environment

One important aspect to refer when coming to development framework, already referred before, is related with capability to adopt a model-based development approach allowing direct support from automatic code generation tools. This is of paramount importance considering that one of the goals of the work is to use the platform to be used by non-experienced users, namely students from high school.

This is the case for the two described software layers, namely the middle layer, which is governed by a state machine and where the goal is to select the proper navigation strategy for each segment. For that, the IOPT-Tools development framework was used [7], allowing automatic generation of C code that can be readily deployed into the Arduino platform. For that end, an IOPT model [8] notation was used to represent the state diagram of Fig. 4.

This is also the case for the lower layer, where the fuzzy controllers were specified using the Xfuzzy framework [9], which allows specification of the set of fuzzy rules, as well as input and output fuzzy sets. Xfuzzy allows generation of C code, as well as C++, Java, and VHDL [10, 11], which after some manual changes can be integrated with the code generated by the IOPT-Tools and directly deployed into the Arduino platform.

Both development frameworks are publically available, free of charge.

4 Improving Electronic Instrumentation and Testing

Experimental results using the presented controllers' characterization proved to be adequate to accomplish the desired control strategies. The presented sailboat controller has been validated in several frameworks, including laboratory as well as navigation in small lakes, and larger and open areas (as in river Tagus, Fig. 8).

However, some weaknesses were identified along experimentation. First of all, the need to have some simulation/emulation environment allowing the developers to test the project even with no navigation conditions, namely in the laboratory. From the three instruments on use, only the wind vane can be operated inside the laboratory forcing any desired value by hand. The GPS receiver is completely useless inside the building. On the other hand, values provided by the compass can be partially tuned (namely different values for the roll and pitch can be read).

Another important weakness is associated with the lack of information about sailboat velocity, as well as wind speed and direction. These two aspects receive special attention, briefly reported in the following subsections.

4.1 Introducing Emulation Environment

A basic emulator, also based in an Arduino board, was developed allowing two types of operation:

Fig. 8 Sailboat in autonomous navigation in river Tagus

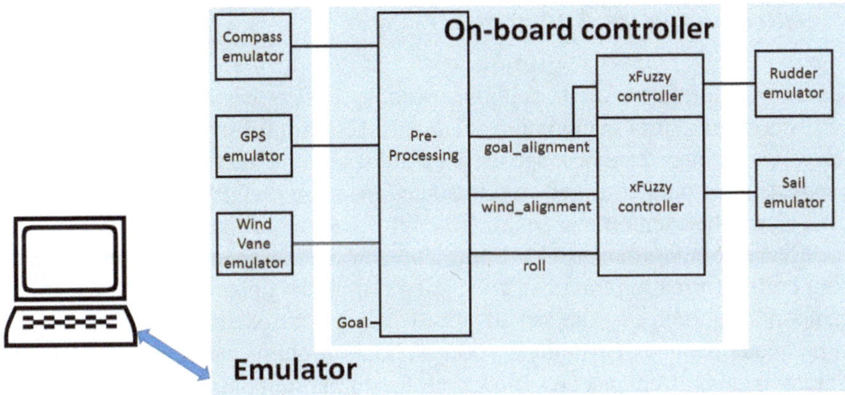

Fig. 9 Emulator framework

- to support laboratory development, where the emulator replaces all the instrumentation devices and communicate with a working station, which can force sequence of values at on-board controller inputs and capture controller's response. This is illustrated in Fig. 9;
- to support data acquisition and control in real navigation situations, where the emulator is connected through a wireless connection to a base station (currently, this option is still work in progress).

For that end, the different interface protocols in use were successfully emulated, being the emulation of the NMEA 0183 - Standard for interfacing marine electronic devices protocol used for GPS receiver [12, 13] the most challenging one (NMEA stands for National Marine Electronics Association).

Some applications and tools are already available to assist autonomous sailboats or other autonomous vehicles. Among them, special mention to METASail [14] and to Neptus [15].

METASail [14] is an application to plan, analyze and supervise the FAST Project [3]. It offers a mission planning environment, where all the path and strategy can be created interactively, by dragging and dropping way-points and assigning or editing actions to them, or even by loading a file with all the instructions. Simulating missions is another functionality of this tool. The tool is also used as an interface to operate and remotely supervise the evolution of the mission, in real operation or simulated.

Neptus [15] was developed within the scope of a network of vehicle systems. The network is composed by human operators, heterogeneous autonomous vehicles and other sensing devices. Neptus platform is used to plan, simulate, monitor and analyze executed missions.

4.2 Introducing an Anemometer

The lack of information about sailboat speed, as well as wind speed and direction, severely constraint the development of advanced control strategies. To mitigate these weaknesses an integrated new sensor is under development. The sensor integrates a homemade anemometer, based on ultrasonic sensors, with GPS and compass instruments, where the anemometer can provide information on wind speed and direction, and GPS and compass can provide information on sailboat speed and direction. Combining both information, one can conclude on true and apparent wind, which is of paramount importance to support advanced sailing control strategies.

The anemometer is currently in final validation phase, using a wind tunnel to calibrate the measurements and allow their use in the sailboat, and will integrate the wind vane previously in use (allowing integration of information from both devices). Preliminary results are very encouraging in terms of accuracy, as well as in terms of cost, where an Arduino Nano is used to assure dedicated computing power.

The ultrasonic based anemometer operates on the measurement of the time of flight of the ultrasonic wave between two ultrasonic transducers separated by approximately 20 cm, as in [16]. The time of flight will depend on the wind speed and direction.

Two pairs of ultrasonic transducers are used, using orthogonal directions, which will allow to determine not only the wind speed as well the wind direction, as in [17]. The structure to support the transducers is lightweight in order not to compromise stability and aerodynamics.

In this sense, it is possible to improve the lower layer of the previously presented controller, associated with the sail and rudder fuzzy controllers, as presented in Fig. 10.

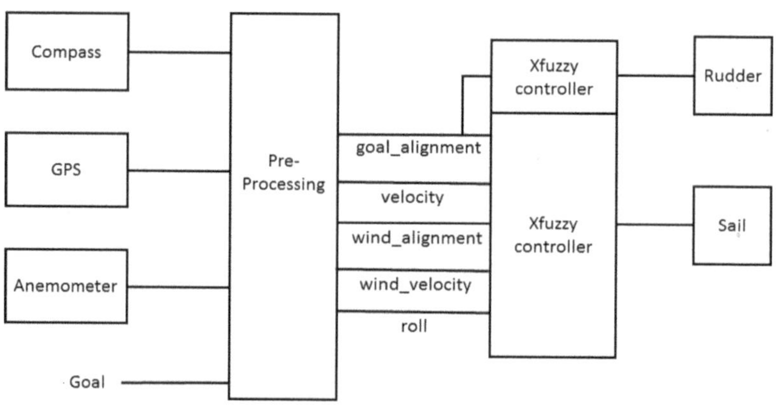

Fig. 10 Block diagram of the new low level controller

5 Conclusions and Future Work

The presented work aims to contribute to have an open platform for autonomous sailboat navigation experimentation, with hardware and software open to the community, as well as to validate model-based development approach to support associated controllers development. In the proposed controller development, two freely available tools frameworks (Xfuzzy and IOPT-Tools) were used allowing automatic C code generation, ready to be deployed in implementation platforms, as Arduino boards. In this sense, development activities will intrinsically take advantage of the usage of simulators and automatic code generators, having a strong impact both at the development time, as well as on the robustness of the solution.

Improvements on electronic instrumentation will continue in order to support more flexibility and accurate control, allowing detection of adverse navigation situations, as well as obstacle detection, and adoption of countermeasures.

Several types of activities have been carried on based on the development of the sailboat controller, namely as support for several MSc Thesis works, and more remarkably being the subject to offer several short Summer courses (one week long) for high school students, as presented in [18], supported by Portuguese Agency Ciência Viva (editions on July 2014, 2015, and 2016). The main goal of the courses is to attract high-school students to sciences and engineering, introducing them to research and development activities within STEM (science, technology, engineering and mathematics) areas, through some exploratory and demonstration activities. Autonomous sailboats control has proved to be an ideal topic to reach this objective.

Future works include the addition of wireless communication allowing exchange of information and collaboration with other sailboats as well as other external agents, within a cyber-physical systems interaction framework.

Acknowledgments This work was partially financed by Portuguese Agency FCT - Fundação para a Ciência e Tecnologia, in the framework of project UID/EEA/00066/2013.

References

1. Arduino home page. http://www.arduino.cc/
2. Alves, J.C., Cruz, N.A.: An FPGA-based embedded system for a sailing robot. In: Proceedings of the EUROMICRO Conference on Digital System Design, Patras, Greece, August 2009 (2009)
3. Alves, J., Cruz, N.: Fast - an autonomous sailing platform for oceanographic missions. In: Proceedings of the MTS-IEEE Conference Oceans' 2008, Quebec, Canada, September 2008. IEEE (2008)
4. Gomes, L., Santos, M., Pereira, T., Costa, A.: Model-based development of an autonomous sailing Yacht controller. In: 2015 IEEE International Conference on Autonomous Robot Systems and Competitions (ICARSC), Vila Real, 2015, pp. 103–108 (2015). doi:10.1109/ICARSC.2015.20

5. Santos, D., Silva Junior, A., Negreiros, A., Vilas Boas, J., Alvarez, J., Araujo, A., Aroca, R., Gonçalves, L.: Design and implementation of a control system for a sailboat robot. In: Robotics, vol. 5(1) (2016). doi:10.3390/robotics5010005
6. Stelzer, R., Proll, T., John, R.I.: Fuzzy logic control system for autonomous sailboats. In: 2007 IEEE International Fuzzy Systems Conference (2007). doi:10.1109/FUZZY.2007.4295347
7. Pererira, F., Moutinho, F., Gomes, L.: IOPT-Tools - Towards cloud design automation of digital controllers with Petri nets. In: 2014 International Conference on Mechatronics and Control (ICMC) (2014)
8. Gomes, L., Barros, J., Costa, A., Nunes, R.: The input-output place-transition petri net class and associated tools. In: 2007 5th IEEE International Conference on Industrial Informatics, June 2007, vol. 1, pp. 509–514 (2007)
9. Fuzzy logic design tools. www2.imse-cnm.csic.es/Xfuzzy/
10. Sanchez-Solano, S., del Toro, E., Brox, M., Brox, P., Baturone, I.: Model-based design methodology for rapid development of fuzzy controllers on FPGAs. IEEE Trans. Indus. Inf. **9**, 1361–1370 (2012)
11. Brox, M., Sanchez-Solano, S., del Toro, E., Brox, P., Moreno-Velo, F.J.: Cad tools for hardware implementation of embedded fuzzy systems on FPGAs. IEEE Trans. Indus. Inf. **60**, 3182–3194 (2012)
12. NMEA, N.: NMEA 0183 Version 4.00 Standard for interfacing marine electronic devices (2008)
13. Langley, R.B.: NMEA 0183: a GPS receiver interface standard. In: GPS World, July 1995, pp. 54–57 (1995)
14. Alves, J.C., Cruz, N.A.: METASail a tool for planning, supervision and analysis of robotic sailboat missions. In: 7th International Robotic Sailing Conference Robotic Sailing 2014, pp. 57–64 (2014)
15. Pinto, J., Calado, P., Braga, J., Dias, P., Martins, R., Marques, E., Sousa, J.B.: Implementation of a control architecture for networked vehicle systems. In: IFAC Workshop on Navigation, Guidance and Control of Underwater Vehicles, pp. 100–105 (2012)
16. Dong, H., Jun, Y.: High accuracy time of flight measurement for ultrasonic anemometer applications. In: 2013 Third International Conference on Instrumentation, Measurement, Computer, Communication and Control (IMCCC) (2013). doi:10.1109/IMCCC.2013.21
17. Xie, K., Wang, K.: Measurement of wind speed and direction with ultrasonic sensor using FPGA. In: Energy Procedia, vol.12, pp.837–843 (2011). doi:10.1016/j.egypro.2011.10.110
18. Gomes, L., Costa, A., Moutinho, F., Mota, R.: Attracting students to engineering through autonomous sailing yatch development. In: ICIT2015 2015 IEEE International Conference on Industrial Technology (2015)

Complex Robot Behavior Creation Using Vector Fields

Alaa El Jawad, Benoît Raymond, Emmanuel Rouault and Fabrice Le Bars

Abstract In term of motion planning for robots, several solutions are possible: grid-based search [1], interval-based search [2], geometric algorithms and potential fields [3]. However, potential fields offer a computational efficient way to generate a desired behavior for robots. However, a principal limit of potential fields is that they deal only with repulsion and attraction. To extend it to perpendicular, tangential and uniform fields, we base our paper on the works of S. Schmitt [4] and R. Arkin [3]. Our contribution consist in developing a method to construct complex vector fields, which are a linear combination of primitive fields. It also explains how to implement this method on a robot using the middleware ROS (Robot Operating System) with any controller.

1 Introduction

Exploring oceans is a big challenge because missions are usually long and repetitive. The relevance of using autonomous sailboats for this kind of missions has already been demonstrated [5], particularly in terms of energy efficiency. One of the objectives of this kind of robot is to avoid collisions with other floating objects in the ocean by detecting them and adapt its behavior. The detection of moving obstacles in the ocean is possible using cameras, laser and/or Automatic Identification System (AIS). The vector field method described below is a solution to take into account these obstacles in the global behavior of the boat and as is based on the work of [4]. In Sect. 2, we focus on the construction of vector fields, then, different applications

A.E. Jawad · B. Raymond · E. Rouault · F.L. Bars (✉)
École Nationale Supérieure de Techniques Avanceés (ENSTA), Bretagne, France
e-mail: alaa.el_jawad@ensta-bretagne.org

B. Raymond
e-mail: benoit.raymond@ensta-bretagne.org

E. Rouault
e-mail: manu.rouault@gmail.com

F.L. Bars
e-mail: fabrice.le_bars@ensta-bretagne.org

J.C. Alves and N.A. Cruz (eds.), *Robotic Sailing 2016*,
DOI 10.1007/978-3-319-45453-5_5

and complex fields are presented in Sect. 3. Finally, in Sect. 4, we talk about the advantages and drawbacks that we experienced with our vector fields and present how we intend to adapt our work on a sailing boat for the WRSC competition.

2 Construction Method

In a vector field there is an assignment of a vector to each point. In the case of this application a point (x;y) is mapped to a vector **f**(x,y). This is a function **f** of the form **f**: $\mathbb{R}^2 \rightarrow \mathbb{R}^2$. For the control method, each point (x,y) corresponds to the robot position, and the mapped vector f(x,y) to the desired behavior composed of informations about:

- the desired speed $\| f(x, y) \|$
- and the heading $\tan^{-1}(\frac{y}{x})$

of the destination's location, so the vector field describes the desired behaviour of the marine robot. Every behavior has two common parameters: the direction it should be heading to and how to approach it (moving away or towards, how fast, and an effect range). So it is interesting to make a distinction between both.

2.1 Field Direction

We can deal with the behavior's location by generating for every point of the space, a vector oriented in the direction of the objective. This is done by some simple geometry depending on the kind of objective (point, line, circle, segment, etc.) Thus, the direction methods in the ***vectorFieldLib.py*** module will generate a vector field corresponding to the chosen objective (Fig. 1).

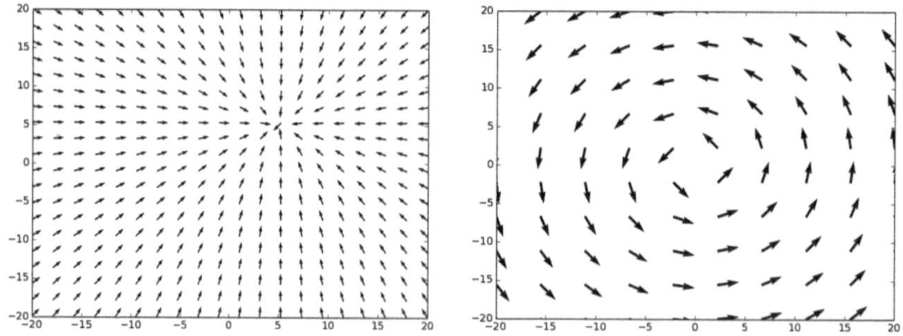

Fig. 1 Attractive point (*left*) and Rotating field (*right*)

It is then useful to deal with the intensity of the generated field, in order to modulate the range, the area of effect and the attraction or repulsion force.

2.2 Field Modulation

A profile is a function defining the intensity of the field relatively to the distance to the objective. Therefore, a profile takes in as parameters:

1. a function that calculates the distances to an objective (i.e. the function calculating a distance to a point is not the same as the one for a segment)
2. function that has the intensity profile or the modulation.

2.3 Atomic Field Addition

One of the advantages of vector fields is that they can be manipulated separately, adjusted to reflect fully the reality and, finally, added, creating a total vector field that has all the underlying behaviors. This implementation has been done by using the python function__add__():

Code 1: __add__ method of the Behavior object

```
def __add__(self, other):
    summed_behavior = Behavior()
    summed_behavior.f_type = ','.join([self.f_type,
        other.f_type])
    summed_behavior.behavior_id = ','.join([self.
        behavior_id, other.behavior_id])
    summed_behavior.cmd_point = lambda x, y: self.
        cmd_point(x, y) + other.cmd_point(x, y)
return summed_behavior
```

Behaviors are implemented as objects. They have a method **cmd_point(x,y)** that maps the point (x,y) which has the desired behavior. In addition, behavior object can be added, and the **summed_behavior(x,y)** must also have a function **cmd_point** which returns a result of the sum of the underlying behaviors.

2.4 Defined Behaviors

Various modulations and summations can be imagined to create more complex vector fields. For example, a circular patrol around a point **P** can be constructed using two atomic fields :

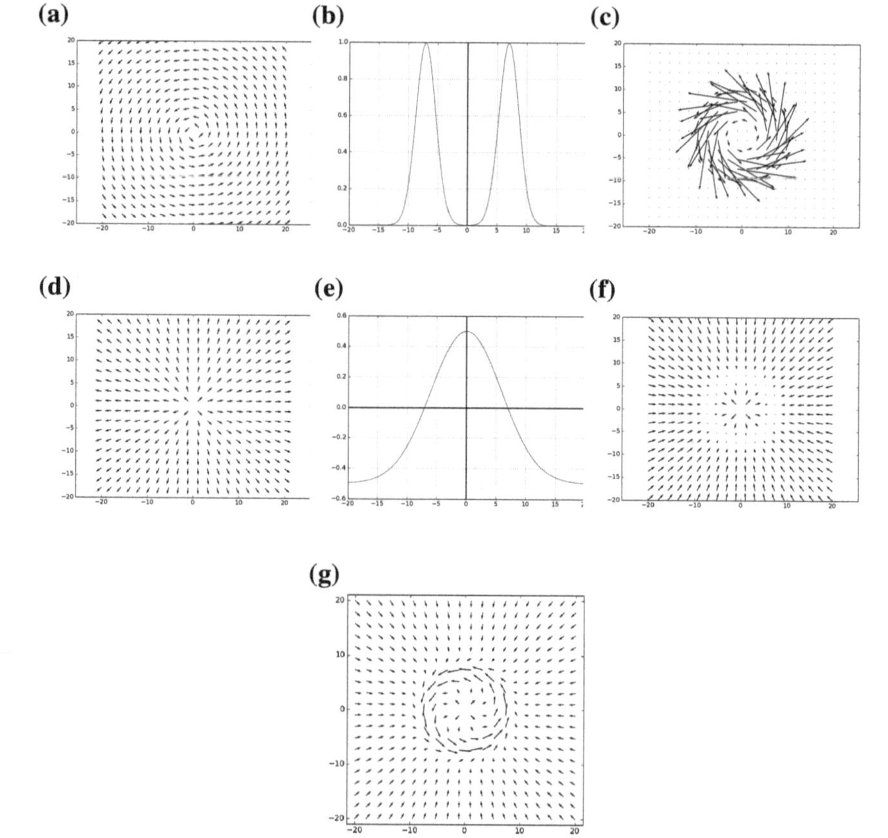

Fig. 2 Patrol field construction

- A rotating (Fig. 2a) which is modulated by a gaussian function (Fig. 2b) centered at a distance r will result in the field of Fig. 2c.
- A point directed field (Fig. 2d) which is modulated by another gaussian (Fig. 2e) centered on the center point **P** that will nullify the field at a distance r and invert it for d(x,y) > r, resulting in the field shown in Fig. 2f.

These two fields are then added to obtain an attractive circular patrol (Fig. 2g) around **P**.

- **Constant field**: defining a global direction to follow
- **Waypoint**: defining a vector field oriented towards a point
- **Limit**: defines a repulsive segment
- **Line**: defines an attractive line to follow
- **Circular patrol**: defines an attractive circle (Fig. 2)
- **Short-range obstacle**: defines the regular repulsive point.

2.5 *Example with WRSC Missions*

The vector filed approach will be used in the **collision avoidance** contest of the WRSC 2016 [6]. This task be implemented by two or three vector fields:

- **Without obstacle**: a vector field telling the robot to cross the rectangle from one side to the other (Fig. 3);
  The robot will follow the line and stay inside of the rectangle, the walls are optional but force the robot to stay inside. Once crossed the boat will change its direction to come back.
- **With obstacle**: a vector field telling the robot to avoid the obstacle, by rotating around it (Fig. 4);
  It is easier than having to generate a list of waypoints that go around the obstacle since no manager is needed.

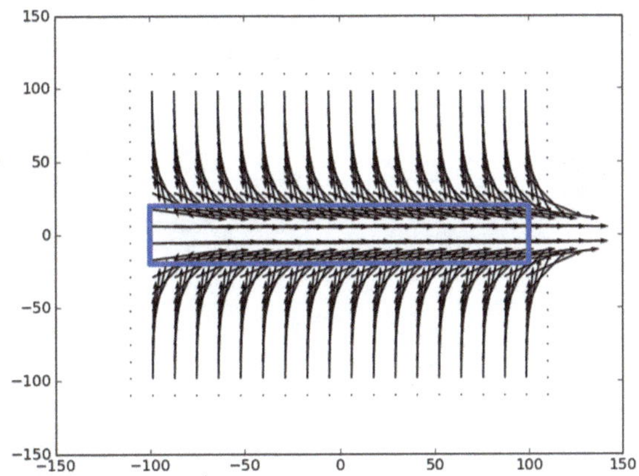

Fig. 3 Vector field line following

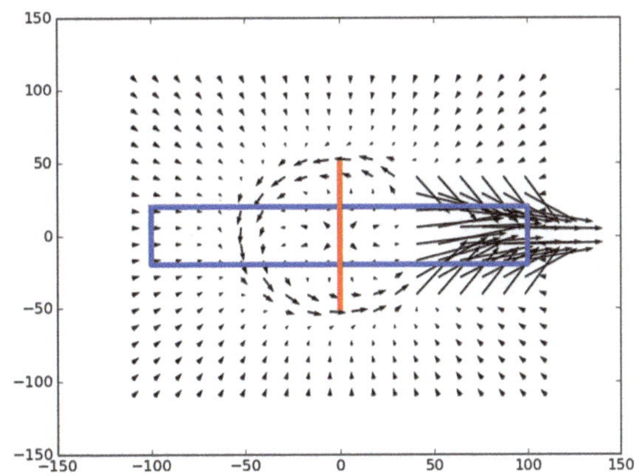

Fig. 4 Vector field with a circular patrol around the obstacle, followed by a line following

3 Implementation

3.1 Architecture: ROS and Python - Flexibility/Adaptability

To implement this vector field method the middleware **ROS** and the programming language **Python** were used. ROS allow to separate the applications. It particularly allow to parallelise easily the processes to take benefit of the *n* cores of processor. It has a publish-subscribe architecture and OS-like advanced functionalities [7]. The node architecture chosen can be seen in Fig. 5.

The communication protocol between ROS nodes allows to refresh the global vector field each time a new obstacle/objective is discovered. The main task is done by the node behavior manager: it manages the behavior list, adding behaviors when new obstacles are detected, or removing them when an objective is achieved. Then, depending on the position of the boat, the node behavior server sends the vector to follow to the field_controller, which translates it as an input for the actuators. The control architecture is also describe (Fig. 6).

For the simulation we used this controller that works on a motor boat:

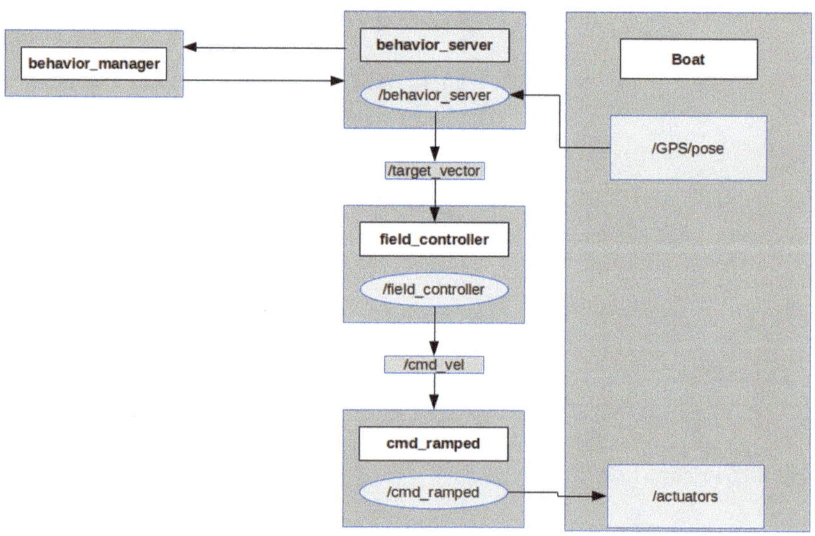

Fig. 5 ROS architecture

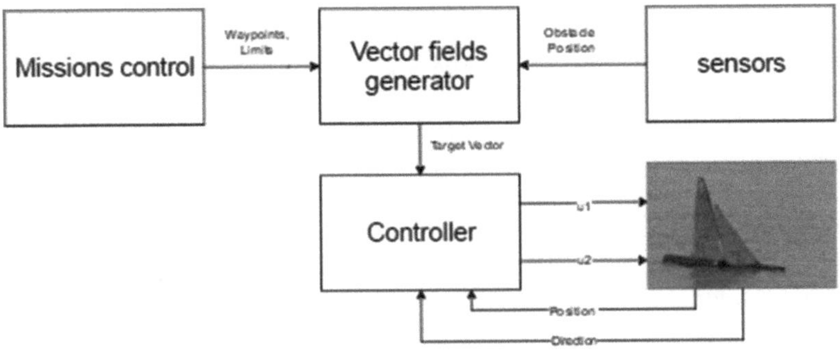

Fig. 6 Control architecture

Code 2: heading controller

```
error = desired_heading-heading # sawtooth function with
    an offset
cmd.angular.z = speed_zero + K * np.arctan(np.tan((error
    / 2.)))

# boat in the direction of the desired heading
if cos(error) >= 0:
    # we scale the command between vLow and vHigh
    if vLow > target_speed:
        cmd.linear.x = vLow
    elif target_speed > vHigh:
        cmd.linear.x = vHigh
    else:
        cmd.linear.x = target_speed
else:
    # reverse is the limit at which the boat needs
    # to reverse gear for security
    if target_speed >= reverse:
        cmd.linear.x = -vHigh
        cmd.angular.z = speed_zero - K * np.arctan(np.
            tan((error / 2.)))
    else:
        cmd.linear.x = vLow
```

with *vLow* and *vHigh* being the minimum and maximum forward speed of the boat.

3.2 *Test and Simulations*

These algorithms were tested first on a ground robot called buggy. However it has a different behavior, since the buggy uses differential driving and can spin around its vertical axis. Therefore, to come closer to the model of the motor boat (or even the sailboat) it is possible to specify a fixed speed to the buggy. First, tests were done by giving to the buggy a waypoint to follow and obtain a very good result (Fig. 7). The buggy correctly reached the waypoint, went through (since it has a constant speed), and then, while it kept trying to reach it, turning around the waypoint. This shows that the waypoint vector field can be also used as a station keeping vector field.

However, testing on a robot requires preparation for every test. In order to make it iterate faster and test more advanced algorithms, we decided to move the testing to a simulation under MORSE (Fig. 8).

Different water environments were created to test all the different behaviors (wall or point avoidance, line following, waypoints, etc.), and a 3D boat that follows a simple model.

All these tests can be easily adapted to sailboats by replacing this simple model by a sailboat model similar to the one presented in [8].

In the simulation (Fig. 8) the environment is composed of three cylinders marking three waypoints and a sphere representing an obstacle. A vector field (Fig. 9) is generated to match to the environment. The boat is placed at different positions and has to reach one of the waypoints while avoiding the obstacle (in this case the target waypoint is the rightmost cylinder). Iterations allow us to find a good vector field and have a good behavior.

Since a motor boat cannot spin around its vertical axis (i.e. it needs a linear speed to turn), when the boat enters an area where the vector field is opposed to the boat current's heading, then the vector field doesn't describe well the behavior. The boat needs a minimum distance to manoeuvre and make a U-turn. This has been corrected

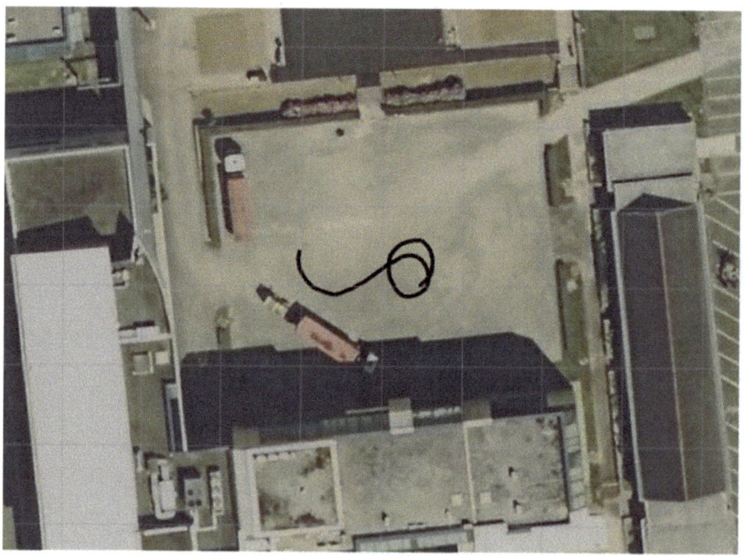

Fig. 7 Trajectory of the buggy

Fig. 8 Morse simulation

in the regulator's described part 3.1 and larger effect range are therefore necessary for obstacles.

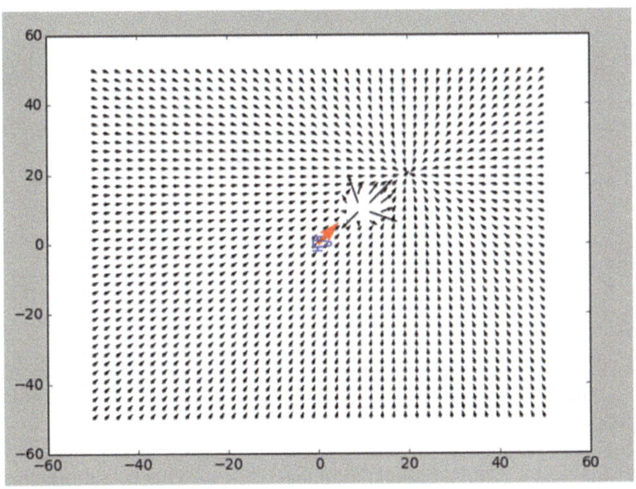

Fig. 9 Morse simulation

4 Ameliorations and Discussions

While vector fields have been used for a while, they are still a very elegant and fast technique for navigation. However, they have some limitations. As described in the previous section, the vector field describes the desired behavior and not the real one. In addition, robots can find themselves in situation where they are stuck, because vector fields confined them in an area which is a local minima. To solve this problem, we can use the current vector field as a base to construct a path, using common algorithms such as A*, or Dijkstra. Our implementation [9] is based on the concept of behaviors, so each behavior generates a vector field. This is different from doing a 2D histogram. In comparison, while 2D histogram is faster to generate vector fields, they cannot handle some kind of behaviors such as rotating field, or event constant field (current flow). On the other side, our implementation requires more thinking and thus more calculation to create complex vector field. However our implementation is very flexible due to using both Python and ROS. So these two approaches can be combined.

References

1. Marija, S., Ivan, P.: Dynamic window based approach to mobile robot motion control in the presence of moving obstacles. In: Proceedings 2007 IEEE International Conference on Robotics and Automation, pp. 1986–1991 (2007)
2. Jaulin, L., Godon, A.: Motion planning using interval analysis. MISC99 Work-shop on Application of Interval Analysis to System and Control (1999)
3. Arkin, R.C.: Motor schema based navigation for a mobile robot : an approach to programming by behavior, vol. 4, pp. 264–271 (1987)
4. Silke, S., Fabrice, L.B., Luc, J., Thomas, L.: Obstacle avoidance for an autonomous marine robot - a vector field approach (2013)
5. Olivier, M., Aymeric, B., Patrick, R., SÃl'bastien, P.: Vaimos: realisation of an autonomous robotic sailboat. In: Proceedings of the 6th International Robotic Sailing Conference, pp. 25–36 (2013)
6. World robotic sailing championship 2016 : Notice of race and preliminary version of rules (2016)
7. Morgan, Q., Brian, G., William, D.S.: Programming Robot with ROS : A Practical Introduction to the Robot Operating System. O'Reilly, California (2016)
8. Fabrice, L.B., Luc, J.: An experimental validation of a robust controller with the vaimos autonomous sailboat. In: Proceedings of the 5th International Robotic Sailing Conference (2012)
9. Alaa, E.J., Benoit, R., Emmanuel, R.: Brest2016, https://github.com/ENSTABretagneRobotics/Brest2016, 2016, [accessed 29-June-2016]

Part III
Sensors and Algorithms for Autonomous Sailing

A Virtual Wind Sensor Based on a Particle Filter

J. Cabrera-Gámez, A.C. Domínguez-Brito, J.D. Hernández-Sosa, B. Valle-Fernández, A. Ramos-de-Miguel and J.C. García

Abstract Wind sensors are essential components of any sailboat, meanwhile they are also one of its most compromised and exposed elements. This paper introduces a novel approach that allows to estimate wind direction and speed based on the application of a particle filter technique that relies on a model dynamics of the sailboat. The proposal incorporates elitism and particle re-initialization to improve filter convergence. Extensive simulation results prove that this approach is capable of providing acceptable estimates of wind conditions at a modest computational cost.

1 Introduction

Wind sensors are key elements for articulating the control strategy of any sailboat, meanwhile they are one of the most exposed components of the ship. This problem is exacerbated in the case of autonomous sailboats where the loss or malfunctioning of a wind sensor can not be fixed or may go unnoticed for a long period of time. It

J. Cabrera-Gámez (✉) · A.C. Domínguez-Brito · J.D. Hernández-Sosa
Instituto Universitario SIANI and Departamento de Informática Y Sistemas,
Universidad de Las Palmas de Gran Canaria, Las Palmas, Spain
e-mail: jorge.cabrera@ulpgc.es

A.C. Domínguez-Brito
e-mail: antonio.dominguez@ulpgc.es

J. Cabrera-Gámez · A.C. Domínguez-Brito · J.D. Hernández-Sosa · B. Valle-Fernández ·
A. Ramos-de-Miguel · J.C. García
Asociación para el Desarrollo de Sistemas Marinos Autónomos—ADSMA,
Las Palmas de Gran Canaria, Spain

B. Valle-Fernández
email: bvallef64@gmail.com

A. Ramos-de-Miguel
email: aramosgcc@gmail.com

J.C. García
email: juancarlos.feroher@gmail.com

J.C. Alves and N.A. Cruz (eds.), *Robotic Sailing 2016*,
DOI 10.1007/978-3-319-45453-5_6

is not an exaggeration to say that the wind sensor is—probably—one of the most important single points of failure in an autonomous sailboat.

The kind of wind sensor usable on board of autonomous sailboats is normally conditioned by its dimensions, i.e. length-over-all or LOA. Basic solutions, frequently used on smaller vessels, have been built as custom designs based on wind-cup anemometers, most of them based on the popular AS5030 non-contact magnetic sensor from Austria Microsystems [1], or as just wind vanes using the MA3 analog absolute rotation encoder from US Digital [2].

The utilization of wind sensors that lack moving or rotating parts is often a big step forward in terms of robustness. Ultrasonic wind sensors belong to this type and, with a large commercial offer, are nowadays the preferred option on larger LOA vessels. Experimental wind sensors without moving parts have also been proposed with the aim of making them more resilient to adverse weather and sea conditions. A good example is the interesting thermal wind sensor design proposed by T. Barton and M. Alvira [3], where mechanical strength and small size are favored at the cost of a larger power consumption.

However, even these compact wind sensors are not immune and their placement on the top of masts and poles expose them to a significant risk of loss and, consequently, devising navigation control strategies that do not rely absolutely on the availability of a wind sensor have full sense. J. Sliwka et al. [4] propose a wind vane self steering device placed at the bow to steer the boat relative to the wind. K. Xiao et al. [5] describe a controller based on interval calculus that does not require explicit knowledge of wind direction.

In order to avoid the use of an explicit physical wind sensor on board, in this paper we explore the approach of estimating the wind direction and wind speed based on the application of a particle filter technique using the dynamic model of the sailboat, and incorporating elitism and re-initialization processes. Thus, in next section, Sect. 2, we introduce the particle filter technique we have applied. In Sect. 3, it is outlined the dynamic model and the sailboat systems equations we have utilized for applying the filter. Section 4 describes the main results we have obtained from simulated experiments to evaluate the approach. And, finally, in Sect. 5 we end with the conclusions we have collected.

2 Particle Filters as a Framework for Positioning, Navigation and Tracking Problems

Particle filters are usually known as recursive implementations of Monte Carlo based statistical signal processing techniques [6]. In problems where we are faced to estimate the state of systems governed by non-linear models and non Gaussian noise, particle filters constitute an alternative approach in real time applications solved typically with other approaches, like the use of Kalman filter techniques [7]. Moreover, considering its computational cost, they are convenient when the computational resources available at run-time are scarce and the system working rate is not very demanding. Both conditions are typically found in autonomous sailboats, where the control hardware on-board is limited in terms of computational resources, and having system operating working rates in the order of seconds.

In fact, in the literature, the particle filter approach has been already applied successfully to the problem areas of positioning, navigation and tracking of moving objects [8]. A well-known problem with the particle filter approach is a degradation of performance when the dimension of the state to estimate grows. A solution for this degradation of performance is the combination of a Kalman filter approach to estimate the derivatives of the state vector, keeping the position estimated using a particle filter. All in all, in general a low dimension of 2 or 3 for the state vector of the particle filter allows to get to an operative real-time algorithm [9].

In general particle filters can be applied to systems following a non-linear model for the state vector, and a non-linear model for measurements, as expressed in Eqs. 1 and 2.

$$x_{t+1} = f(x_t, u_t) + f_t \quad (1)$$

$$y_t = h(x_t) + e_t \quad (2)$$

Where x_t is the state vector, $f(x_t, u_t)$ is the state vector model, u_t the measurement inputs, $h(x_t)$ the model for the measurements, and f_t and e_t are respectively state vector and measurement errors. It is assumed independent distributions for f_t, e_t and x_0, with known probability densities p_{f_t}, p_{e_t} and p_{x_0}, respectively not necessarily Gaussian. When the model for the state is linear the previous equations become Eqs. 3 and 4.

$$x_{t+1} = Ax_t + Bu_t + B_f f_t \quad (3)$$

$$y_t = h(x_t) + e_t \quad (4)$$

The particle filter approach consist of a numerical implementation to approximate the posterior distribution, $p(x_t|Y_t)$, of the state vector, applying the algorithm shown in Fig. 1 (taken from [9]).

3 Estimating Wind Conditions Using a Particle Filter

Using the particle filter approach described in the former section, it is possible to estimate wind conditions, speed and direction, using a dynamic model of the sailboat. Within this approach, each particle represents an hypothesis about wind conditions and the filter will try to find the set of particles that best explains the motion of the sailboat for a short period of time, under current wind conditions.

1. *Initialization*: Generate $x_0^i \sim p_{x_0}, i = 1, ..., N$. Each sample of the state vector is referred to as a particle.
2. *Measurement update*: Update the weights by the likelihood:

$$w_t^i = w_{t-1}^i p(y_t|x_t^i) = w_{t-1}^i p_{e_t}(y_t - h(x_t^i)), i = 1, 2, ..., N$$

and normalize to $w_t^i \leftarrow \frac{w_t^i}{\sum_i w_t^i}$. And as an approximation to

$$\hat{x}_t = \int x_t p(x_t|Y_t) dx_t$$

take

$$\hat{x}_t \approx \sum_{i=1}^{N} w_t^i x_t^i$$

3. *Re-sampling*: To make the particle population evolve, several re-sampling strategies can be applied. Two of the most commonly used are the followings:

   a. *Bayesian bootstrap*. Take N samples with replacement from the set $\{x_t^i\}_{i=1}^{N}$ where the probability to take sample i is w_t^i. Let $w_t^i = \frac{1}{N}$. This step is also called Sample Importance Re-sampling (SIR).
   b. *Importance sampling*. Only re-sample as above when the effective number of samples is less than a threshold N_{th},

$$N_{eff} = \frac{1}{\sum_i (w_t^i)^2} < N_{th}$$

   Here $1 \leq N_{eff} \leq N$, where the upper bound is attained when all particles have the same weight, and the lower bound when all probability mass is at one particle. The threshold can be chosen at $N_{th} = 2\frac{N}{3}$.

4. *Prediction*: Take a $f_t^i \sim p_{f_t}$ and simulate:

$$x_{t+1}^i = f(x_t^i, u_t^i) + f_t, i = 1, 2, ..., N$$

5. Let $t \leftarrow t+1$ and iterate to step 2.

Fig. 1 Particle filter algorithm

In order to predict the motion of particles according to hypothesized wind conditions a sailboat dynamic characterization is needed. The model used in our simulations has been adapted from [10], and includes two control inputs (rudder and sail positions), two external inputs (wind direction and speed) and ten configuration parameters. As a result, the sailboat description is summarized here by the following state space equations (see the original reference for details):

$$\begin{cases} \dot{x} = v\cos(\theta) + p_1 a_{tw}\cos(\psi_{tw}) \\ \dot{y} = v\sin(\theta) + p_1 a_{tw}\sin(\psi_{tw}) \\ \dot{\theta} = \omega \\ \dot{v} = \frac{f_s\sin(\delta_s) - f_r\sin(u_1) - p_2 v^2}{p_9} \\ \dot{\omega} = \frac{f_s(p_6 - p_7\cos(\delta_s)) - p_8 f_r\cos(u_1) - p_3\omega v}{p_{10}} \end{cases} \tag{5}$$

$$\begin{cases} \boldsymbol{w}_{aw} = \begin{pmatrix} a_{tw}\cos(\psi_{tw}-\theta)-v \\ a_{tw}\sin(\psi_{tw}-\theta) \end{pmatrix} \\ \psi_{aw} = atan2(\boldsymbol{w}_{aw}) \\ a_{aw} = \|\boldsymbol{w}_{aw}\| \\ \gamma = \cos(l)+\cos(\psi_{aw}) \\ l = |\delta_s| = f(u_2) \quad if \quad \gamma > 0 \\ \delta_s = \begin{cases} -arctan(\tan(\psi_{aw})) & if \quad \gamma \leq 0 \\ -l sign(\sin(\psi_{aw})) & otherwise \end{cases} \\ f_s = p_4 a_{aw}\sin(\delta_s - \psi_{aw}) \\ f_r = p_5 v \sin(u_1) \end{cases} \tag{6}$$

Where $\dot{x}$ and $\dot{y}$ represent the horizontal velocity components in North and East directions, respectively; θ is the heading relative to the North, and $\dot{\theta}$ is the angular velocity of the sailboat; v is the tangential speed; f_s is the lift force due to the sail and f_r is the lift force on the rudder; $\boldsymbol{w}_{aw}$ is the apparent wind vector; a_{tw} and a_{aw} represent the true wind and the apparent wind speed, respectively; similarly, ψ_{tw} and ψ_{aw} are the corresponding true and apparent wind directions. Note that in these equations wind direction refers to direction of flow. In fact, if $\tilde{\psi}_{aw}$ is the apparent wind angle that would be reported by a wind sensor, i.e. the wind incidence angle, $\psi_{aw} = fmod(\tilde{\psi}_{aw} + pi, 2\pi)$. All angles are considered positive clockwise.

Regarding the model inputs, $u_1 = \delta_r$ and $u_2 = |\delta_s|$ are the control variables, where u_1 represents the rudder angle relative to boat's main axis; and u_2 is the mainsail's sheet length. l is the pretended or potential mainsail aperture, a function of the sheet length. The environmental inputs a_{tw} and ψ_{tw} represent the absolute or true wind speed and direction, respectively.

The model configuration parameters, p_i, are assumed to be known: p_1 is the drift coefficient; p_2 and p_3 represent, respectively, the tangential and angular frictions; p_4 is the sail lift; p_5 is the rudder lift; p_6, p_7 and p_8 are geometrical coefficients of the sailboat (see Fig. 2); p_9 is the mass of the boat and p_{10} its mass moment of inertia.

In this work we have used the parameter set described in [10], $p_1 = 0.05$, $p_2 = 0.2\,\text{kg/s}$, $p_3 = 6000\,\text{kg·m}$, $p_4 = 1000\,\text{kg/s}$, $p_5 = 2000\,\text{kg/s}$, $p_6 = 1\,\text{m}$, $p_7 = 1\,\text{m}$, $p_8 = 2\,\text{m}$, $p_9 = 300\,\text{kg}$, $p_{10} = 10000\,\text{kg·m}^2$, length over all (LOA) = 3.65 m.

Displacement vessels have their attainable speed, v, limited by its waterline length (LWL) [11] due to the motion-induced wave. The so called hull velocity, v_{hull} can be computed approximately as:

$$v_{hull}(knots) \approx 2.43\sqrt{LWL}$$

Where the LWL is given in meters. This aspect has been incorporated to the model to limit the maximum sailboat's speed.

Fig. 2 Sailboat model

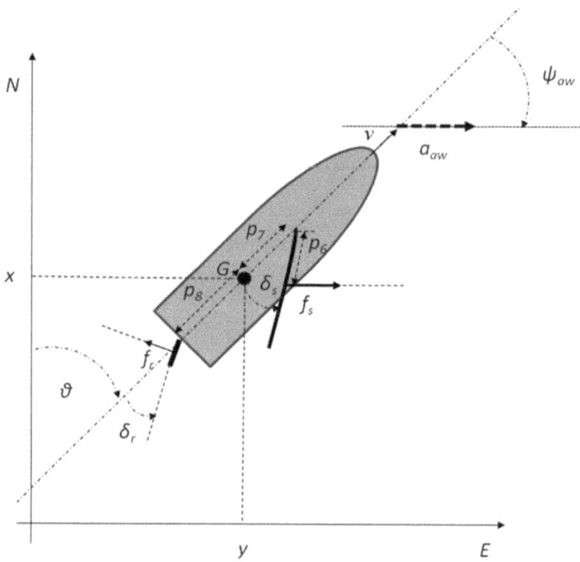

4 Results

A series of simulations have been performed in Matlab to test the validity of the proposed approach. The experimental setup has been defined for a short term navigation problem using a time step of 100 ms and a total simulation time of 5 min, as a convenient balance between the resolution of the dynamic model and the processing requirements.

Considering the limited computational resources that will be available aboard, the state dimension of the particle filter has been limited to 2: the estimation of the direction and the speed of the wind.

The environmental conditions used for testing consisted in 16 equidistant wind directions (separated by 22.5 ° increments) and three window speeds (1.75, 3.5 and 7 m/s), for a total of 48 test scenarios. In all simulations, both the rudder angle and the mainsail angle, have been kept constant ($\delta_r = 0$, $\delta_s = 45$ °).

After some preliminary analyses, a base case has been configured for the filter with 20 particles and 5 s of cycle time. The filter initialization of the particles selects random samples from a Uniform PDF (Probability Density Function) for the wind direction between 180 ° and 180 °, while the wind speed is randomly sampled from a Weibull PDF with a 2 shape factor and a 10 scale value. The observation function for the particle weighting is based on the euclidean distance between the real sailboat trajectory and the one predicted according to the particle state estimation. The observation error is characterized as a Normal PDF with zero mean, 5 m of typical deviation. The proposed scheme implements best particle elitism and sample importance re-sampling, with additive mutation processes characterized by Normal PDFs with 5° and 0.2 m/s dispersion. A supplementary particle re-initialization process is applied for the 25 % lower weighted ones when no filter convergence is detected.

In the following experiments, a zero mean Gaussian noise has been added to the true wind direction and speed values while simulating the sailboat trajectory, using 5° and 0.5 m/s as typical deviations, respectively. Every scenario has been tested for 51 independent runs. Note that in all graphs true wind incidence angle is used in the horizontal axis.

Figures 3 and 4 show the wind estimation error analysis for the base case. Results show a median absolute direction estimation error generally below 5° and a median absolute speed estimation error generally below 0.25 m/s, with only test directions around $-75°$ and $75°$ degrees showing a worse result.

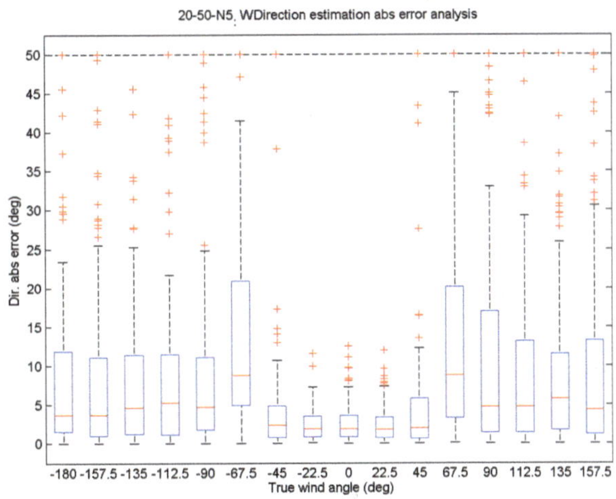

Fig. 3 Box plot for wind direction absolute errors (50° data limit) as a function of the different test directions

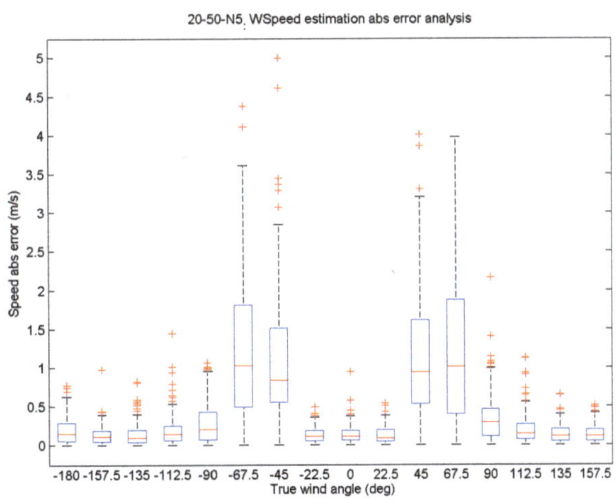

Fig. 4 Box plot for wind speed absolute errors as a function of the different test directions

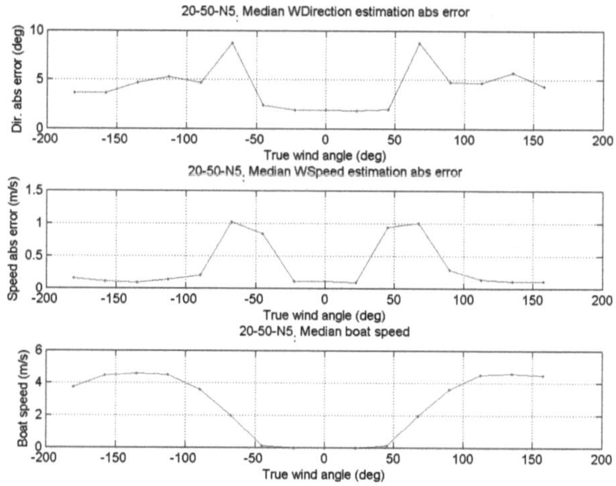

Fig. 5 Wind estimation absolute errors and boat speed as a function of the different test directions

Figure 5 illustrates that the worse estimation results correspond to situations where the boat shows higher variability with the wind direction. The effect of speed saturation is visible in the bottom graph.

The base case has been compared with two alternative configurations: one with double number of particles and another with double cycle interval. Figures 6 and 7 illustrate the performance comparison between all three alternatives in terms of median and standard deviation values. The configuration using 10 s of cycle interval offers slightly better results in the central directions, where the boat speed is low and increasing the integration interval is positive. On the contrary, the configuration using 40 particles performs better off the central region, where the higher boat speeds demands the use of more particles in the filter. Globally, we consider these improvements are not significant enough, and the base configuration seems to be a reasonable choice.

In order to evaluate the effect of elitism selection, the base case simulation has been repeated deactivating this mechanism. The results, focusing in wind direction estimation error, show that elitism contributes with a global improvement of a 52.7 % in the median value and a 29.1 % in the standard deviation. Similarly, regarding the no-convergence re-initialization mechanism, the effect has been evaluated as a global improvement around a 19 % in median value and a 13 % in standard deviation.

Some extreme cases have been also tested for evaluating the filter stability. Using configurations with as low as 5 particles and noise measurement levels of 10 meters the scheme is still able to produce correct median value estimations, though the high dispersion makes them impractical for real time operation, because a high number of wind estimations would need to be averaged to get a reliable result.

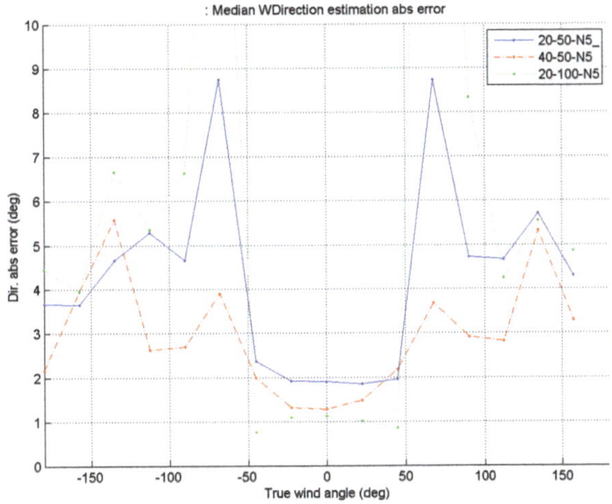

Fig. 6 Performance comparison with two alternative filter configurations: 40 particles and 10-s cycle—median values

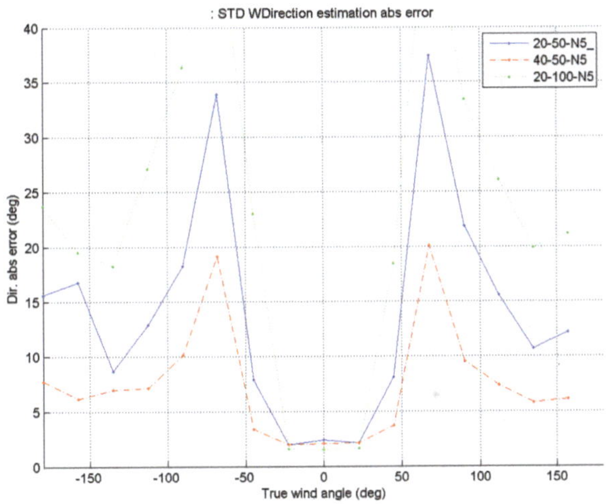

Fig. 7 Performance comparison with two alternative filter configurations: 40 particles and 10-s cycle—standard deviation values

5 Conclusions

This paper has introduced a virtual wind sensor based on a particle filter, incorporating elitism and re-initialization mechanisms. The results achieved in simulation under realistic noise conditions have produced good estimates of wind direction and speed for most points of sailing. The influence of different aspects such as the number of particles and the integration interval have been analyzed, as well as the contribution of the filtering improvement mechanisms. Specially interesting, in order to implement it on the on-board microcontroller, is the fact that this approach still provide reliable estimates with small particle populations.

The primary application of this filter could be to replace the sailboat's wind sensor on board just in case it ceases to operate. But this virtual sensor could be used also to confirm wind measurements reported by the wind sensor available on board.

Future work will address more detailed simulation studies, for example, taking into account leeward and currents effect, and it will be devoted also to testing this virtual wind sensor on a real sailboat.

Acknowledgments The authors are sincerely grateful to Solumatica Canarias for providing financial support for building the A-TIRMA G2 prototype and the Real Club Náutico de Gran Canaria for the access granted to its facilities during the development of this project.

References

1. Neal, M., Sauzé, C., Thomas, B., Alves, J.C.: Technologies for autonomous sailing: wings and wind sensors. In: Proceedings of the 2nd International Robotic Sailing Conference, Robotic Sailing 2009, Matosinhos, Portugal, pp. 23–30, 6–12th July 2009
2. Cabrera-Gámez, J., Ramos de Miguel, A., Domínguez-Brito, A., Hernández-Sosa, J., Isern-González, J.D., Fernández-Perdomo, E.: An embedded low-power control system for autonomous sailboats. In: Bars, F.L., Jaulin, L. (eds.) Proceedings of the 6th International Robotic Sailing Conference, Robotic Sailing 2013, Brest France, September 2013, Springer (2013)
3. Barton, T., Alvira, M.: A discrete-component 2d-array wind sensor without moving parts for a robotic sailboat. In: Proceedings of the 5th International Robotic Sailing Conference, Robotic Sailing 2012, Cardiff, Wales, September 2012, pp. 95–104, Springer (2012)
4. Sliwka, J., Nicola, J., Coquelin, R., de Megille, F.B., Clement, B., Jaulin, L.: Sailing without wind sensor and other hardware and software innovations. In: Schlaefer, A., Blaurock, O. (eds.) Proceedings of the 4th International Robotic Sailing Conference, Robotic Sailing, August 2011, Lubeck, Germany. Springer (2011)
5. Xiao, K., Sliwka, J., Jaulin, L.: A wind-independent control strategy for autonomous sailboats based on voronoi diagram. In: Proceedings of the 14th International Conference on Climbing and Walking Robots and the Support Technologies for Mobile Machines, CLAWAR, University Pierre et Marie Curie, Paris, France, 6–8 September 2011
6. Crisan, D.: Sequential Monte Carlo Methods in Practice. Ch. Particle Filters—A Theoretical Perspective, pp. 17–41. Springer, New York (2001). http://dx.doi.org/10.1007/978-1-4757-3437-9_2
7. Bar Shalom, Y., Li, X.-R., Kirubarajan, T.: Estimation with Applications to Tracking and Navigation. Wiley, New York (2001)
8. Montemerlo, M., Thrun, S.: FastSLAM: A Scalable Method for the Simultaneous Localization and Mapping Problem in Robotics. Springer, Berlin (2007)
9. Gustafsson, F., Gunnarsson, F., Bergman, N., Forssell, U., Jansson, J., Karlsson, R., Nordlund, P.J.: Particle filters for positioning, navigation, and tracking. IEEE Trans. Signal Process. **50**(2), 425–437 (2002)
10. Jaulin, L., Bars, F.L.: Sailboat as a windmill. In: Bars, F.L., Jaulin, L. (eds.) Proceedings of the 6th International Robotic Sailing Conference, Robotic Sailing, September 2013, Brest, France. Springer (2014)
11. Marchaj, C.: Sailing Theory and Practice, 2nd edn. Mead & Company, New York (1982)

Effect of an Ensemble Algorithm in Reinforcement Learning for Garbage-Collection Sailing

Kanta Tachibana and Ryuta Fukazawa

Abstract A robot sailor can obtain its behaviour autonomously with reinforcement learning. However, reinforcement learning suffers from the curse of dimensionality, with an increase in state variables and an exponential increase in the number of states to realize fine control. This paper introduces an ensemble algorithm in Q-learning to allow robot sailors to collect garbage while sailing, and discusses the effect of the ensemble algorithm. This paper especially investigated the enhancement of decision-making to sail faster to the target position, while keeping a small number of state variables and a small number of states. Numerical experiments show a statistically significant enhancement by the proposed ensemble decision-making algorithm with a diverse number of agents, state variables, and learning parameters.

1 Introduction

Reinforcement-learning algorithms [1–3], such as Q-learning, Actor-Critic learning, etc., have been proposed as machine-learning methods to learn behaviour without supervised signals. In reinforcement learning, agents update the value of an action in various states in response to received positive or negative rewards. It is applied to many control problems [4] because it does not need pre-defined control rules and is suitable for time-varying, unsteady environments. However, it easily suffers from the so-called "curse of dimensionality"; i.e., it converges to local minima and the robot eventually repeats undesirable behaviour in large-dimension state spaces [5].

Because robot-control problems in the real world—e.g., inverse pendulum [6], learning-to-stand-up motion [7], and route finding [8]—have a continuous state space, it is important to suitably discretize the state space. Fuzzy Q-learning [9, 10] was proposed to fuzzily discretize a continuous state space. It effectively

K. Tachibana (✉) · R. Fukazawa
Faculty of Informatics, Kogakuin University, Tokyo, Shinjuku 1-24-2, Nishishinjuku, Japan
e-mail: kanta@cc.kogakuin.ac.jp

R. Fukazawa
e-mail: f07r01@gmail.com

J.C. Alves and N.A. Cruz (eds.), *Robotic Sailing 2016*,
DOI 10.1007/978-3-319-45453-5_7

expresses a continuous state and absorbs measurement noise from many sensors. The autonomous reconstruction of state space [11] and an adaptive switching controller [12] were proposed as other discretization methods that are based on reinforcement learning. Learning the rudder control of a sailing yacht [13, 14] is a more difficult problem in the class of partially observable Markov decision processes (POMDPs) because a map of the future wind speed and direction is only available stochastically.

On the other hand, Ensemble Learning was proposed [15] as a strong scheme for supervised learning. Ensemble learning consists of a group of weak learners. Weak learners learn limited supervised data, i.e., a part of a dataset or a subset of explanatory variables. The decisions made by the weak learners are integrated into a single decision through a maximal vote, etc. The integrated decision is shown to be better than one made by a single strong learner that learns all the explanatory variables of the entire dataset. Sugiyama et al. [16] showed the effectiveness of ensemble learning in their computerized Shogi game. Adaboost [17] was proposed to adaptively weigh each weak learner.

Wiering et al. [18] applied an ensemble-learning scheme to reinforcement learning. In their study, weak learners, each of whom learned with either Q-learning, Actor-Critic learning, or Sarsa learning, made different decisions from each other, and the integrated decision was better than one made by a single agent. In their study, each agent had a different learning model and used all available state variables.

This paper proposes an ensemble algorithm in reinforcement learning for garbage-collection sailing. For autonomous sailing, a prompt response is necessary against rapid wind changes. In contrast to the proceeding study [18], all agents learn with Q-learning and each agent has a limited number of state variables, except for one agent.

Section 2 describes the setting of garbage-collection sailing and reinforcement learning. In Sect. 3, we propose an ensemble algorithm in Q-learning. Section 4 presents the results of numerical simulations to evaluate the effect of the proposed ensemble algorithm. Section 5 discusses which agent contributes to the decision-making and Sect. 6 describes our conclusions.

2 Garbage-Collection Sailing

We implemented a simulator for garbage-collection sailing in a randomly shifting wind field. In the simulation, a robotic sailor is trained to collect garbage effectively with a reinforcement learning method. Section 2.1 defines the state variables and Sect. 2.2 describes in detail the reinforcement learning and the action selection.

2.1 Simulation Setting

In our two-dimensional sailing simulations, we assume that sailing yachts are accelerated by a sail-force propulsive component and decelerated by water resistance. The direction of the yacht is controlled by a robot sailor. We assume that the leeway effect, i.e., lateral acceleration, is zero. This assumption means that the lateral resistance force of the water is large enough to cancel the lateral component of the sail force.

We also assume that the wind speed is constant throughout the simulation period, and the wind blows in the same direction at any point in time. We assume that the wind direction periodically changes $\Delta\phi \sim U(-5, 5)$[deg], where $\Delta\phi$ is generated from a uniform distribution within a range of $\pm 5°$. We set the garbage's position as a target point on the ocean, and inform the robot sailor of its exact direction. After the sailing yacht reaches the target position, a new target position is set at random.

We assume that the lateral force given by the wind is totally cancelled out by the lateral resistance force given by the water. We assume that the sail force is proportional to the square of the apparent wind speed, $\|\vec{w}_r\|^2$, and proportional to the sail area facing the wind, $\sin(\varphi - \varphi')$, where φ is the apparent wind direction and φ' is the sail angle. To simplify the learning problem, we assume that the sail angle is always controlled to maximize $\sin\varphi' \sin(\varphi - \varphi')$, which is the propulsive component of the sail force. When the sail angle is $\phi' = \frac{\varphi}{2}$, the propulsive acceleration a_x is calculated as

$$a_x = \frac{\|\vec{w}_r\|^2(1 - \cos\varphi)}{2} - R_x\|\vec{v}\|^2,$$

where $\vec{v}$ is the velocity vector. We set the coefficient of water resistance $R_x = 0.0005$.

On the other hand, the rudder is controlled by either reinforcement or ensemble learning. We assume the yacht direction is holonomic. If the robot sailor selects to go straight, the yacht direction does not change, and if it selects to go right or left, the yacht direction changes exactly ± 3[deg]. Then, the velocity vector is updated as $\vec{v} \leftarrow \left[R_{a[i]}\right]\left\{\vec{v} + \binom{a_x}{0}\right\}$, where $\left[R_{a[i]}\right]$ is a rotation matrix to rotate $a[i] = \pm 3$ or 0[deg]. The yacht position is then updated as $\vec{x} \leftarrow \vec{x} + [R]\vec{v}$, where $[R]$ is the rotation matrix from the yacht coordinate system to the global coordinate system.

2.2 Reinforcement Learning

2.2.1 Q-Learning

Watkins [1] proposed a reinforcement-learning method in which an agent updates a Q-value, which is the evaluation value of a state/action pair corresponding to its sequence of states and actions. The authors proved the convergence of Q-learning after sufficient repetitions. In the case where an agent was in state s_t, took action a_t, and received reward r_t, the Q-value for the pair (s_t, a_t) was updated as:

$$Q(s_t, a_t) \leftarrow (1-\alpha)\, Q(s_t, a_t) + \alpha \left\{ r_t + \gamma \max_a Q(s_{t+1}, a) \right\},$$

where preset parameters $\alpha \in (0, 1]$ and $\gamma \in (0, 1)$ are the learning rate and discount factor, respectively. The last term evaluates the maximal Q-values in the state at the next time-step. Thus, the positive/negative reward is back-propagated along with the good/bad sequence of actions.

Some action-selection policies have been proposed, such as ε-greedy and soft-max. Under the ε-greedy policy, the agent selects the action with the maximal Q-value with probability $(1-\varepsilon)$, and selects one of the possible actions at random with probability ε. Under the soft-max policy, the agent selects action a in state s with the probability:

$$\pi(s, a) = \frac{\exp(Q(s, a)/T)}{\sum_i \exp(Q(s, a[i])/T)},$$

where T is a temperature parameter to control the selection randomness. An agent selects its actions more randomly with a higher T, and selects the action with the largest Q-value more deterministically with a lower T.

2.2.2 Definition of State Variables

Agents can observe at most three variables: s_1, the direction of the target relative to the yacht heading; s_2, the yacht heading relative to the wind; and s_3, the direction of the target relative to the wind. We discretize these variables as follows:

- s_1 : The target position is to the right (starboard side, $s_1 = 1$), to the front ($s_1 = 0$), or to the left (port side, $s_1 = -1$).

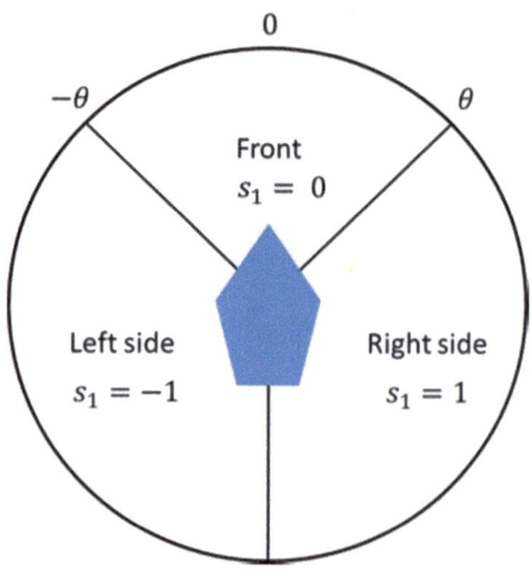

Fig. 1 Target-direction state variable

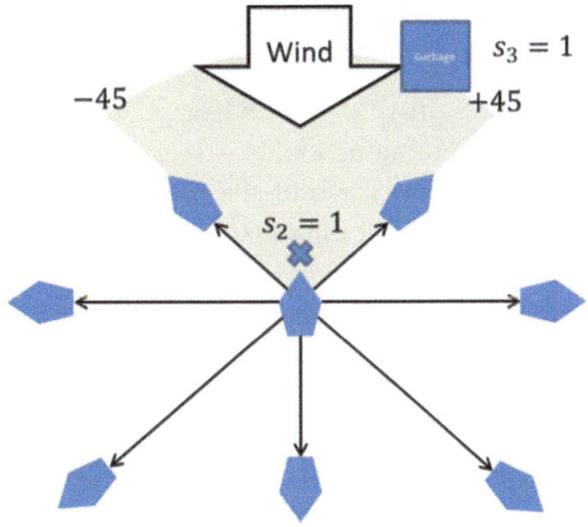

Fig. 2 Dead-zone state variables

- s_2 : The yacht heads into the dead zone ($s_2 = 1$) or not ($s_2 = 0$).
- s_3 : The target position is in the dead zone ($s_3 = 1$) or not ($s_3 = 0$).

To discretize s_1, we use a discretizing parameter θ, as shown in Fig. 1. If the direction of the target position is in the range between θ left and θ right, $s_1 = 0$; i.e., the garbage is to the front. The dead zone is defined as a range between 45° left and 45° right, centred in the wind direction, as shown in Fig. 2.

3 Ensemble Algorithm

An ensemble algorithm integrates the policies of many agents to make a decision. Maximal voting is the easiest way to integrate policies. Wiering et al. [18] investigated methods of integrating many agents' policies. Among maximal voting, Boltzman sum, Boltzman product, weighted sum, and weighted product, they found the best integration was accomplished with the Boltzman product.

Here, let s_t show the state vector at time-step t. The j-th agent has policy π^j, which is a mapping from a state/action pair to a positive real number. The output of π^j is larger for a more likely action in the given state. Thus, $\pi_t^j(s_t, a[i])$ shows the probability that the agent will select action $a[i]$. The Boltzman product is a method of integrating decisions to obtain the ensemble decision $\pi_t(s_t, a[i])$ so that it is proportional to the product of the decisions made by the weak learners:

$$p_t(s_t, a[i]) = \prod_j \pi_t^j(s_t, a[i]),$$

$$\pi_t(s_t, a[i]) = \frac{p_t(s_t, a[i])^{\frac{1}{\tau}}}{\sum_k p_t(s_t, a[k])^{\frac{1}{\tau}}},$$

where τ is a temperature parameter.

3.1 Diversity

The agents are diverse. Each agent uses a different subset of state variables, a different learning parameter, and a different discretization parameter. One agent uses all three state variables. Each of the other agents uses one or two state variables and has a different learning rate, discount factor, and discretization parameter θ.

3.2 Ensemble-Learning Workflow

Figure 3 shows the workflow of the proposed ensemble learning. First, the wind and yacht are initialized, the agents are set to control the yacht, and the garbage is placed in a random position. Then, the agents observe the state. The right side of Fig. 5 shows the workflow of each agent. Each agent generates a virtual yacht at the same position, direction, and speed as the real yacht; then, it selects the action of the virtual yacht following its Q-value, π_t^j $(s_t, a[i])$. If the virtual yacht reaches the garbage position as the result of the selected action, it receives a positive reward. If not, it receives a negative reward. The Q-value is updated according to the state and action. Next, the decisions produced by each agent are integrated and the yacht moves according to the integrated decision.

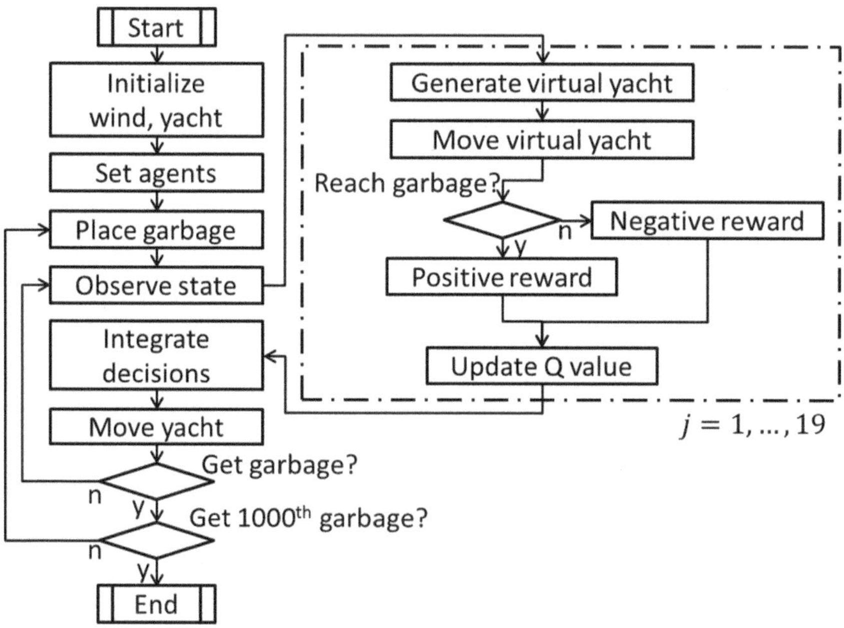

Fig. 3 Ensemble-learning workflow for garbage-collection sailing

4 Numerical Experiments

To clarify the effect of an ensemble algorithm in Q-learning for garbage-collection sailing, two types of experiments were executed. In experiment 1, a single agent learns to collect the garbage in a randomly shifting wind field. We called this learned agent the Alpha-agent. In experiment 2, two yachts compete to reach the garbage position faster. One yacht is controlled by the Alpha-agent. The other yacht is controlled by the integrated decision of 19 agents, including the Alpha-agent.

4.1 Experiment 1: Behaviour Learning by a Single Agent

The Alpha-agent learns garbage-collection behaviour with a soft-max action selection using the following parameter settings: observable state variables $\{s_1, s_2, s_3\}$, $\alpha = 0.001$, $\gamma = 0.999$, $\theta = 45$, and $T = 0.7$. After the Alpha-agent achieves good behaviour, the yacht can collect garbage smoothly, as shown in Fig. 4. Because the sail angle is always controlled to maximize the propulsion and the yacht is not decelerated, it goes in circles until it finally reaches the garbage.

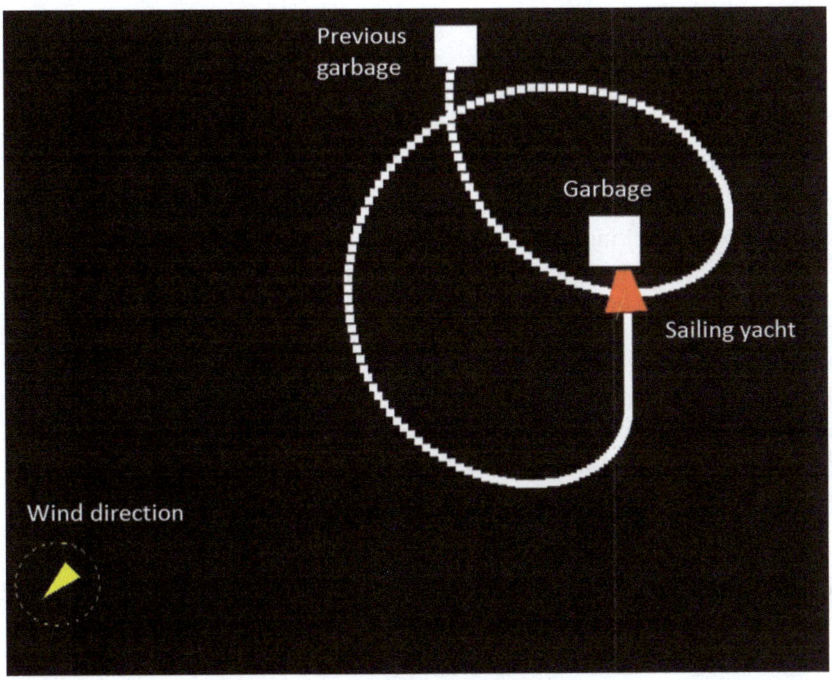

Fig. 4 Alpha-agent's learned garbage-collection behaviour

Table 1 List of agents. Learning rate α, discount factor γ, discrimination parameter θ

ID	α	γ	θ	State variable(s)	Pre-learned
1	0.001	0.999	45	s_1, s_2, s_3	Yes
2	0.01	0.99	45	s_1	No
3	0.01	0.99	–	s_2	No
4	0.01	0.99	–	s_3	No
5	0.01	0.99	45	s_1, s_2	No
6	0.01	0.99	45	s_1, s_3	No
7	0.01	0.99	–	s_2, s_3	No
8	0.02	0.98	44	s_1	No
9	0.02	0.98	–	s_2	No
10	0.02	0.98	–	s_3	No
11	0.02	0.98	44	s_1, s_2	No
12	0.02	0.98	44	s_1, s_3	No
13	0.02	0.98	–	s_2, s_3	No
14	0.03	0.97	43	s_1	No
15	0.03	0.97	–	s_2	No
16	0.03	0.97	–	s_3	No
17	0.03	0.97	43	s_1, s_2	No
18	0.03	0.97	43	s_1, s_3	No
19	0.03	0.97	–	s_2, s_3	No

4.2 Experiment 2: Effect of Ensemble Learning

Garbage-collection sailing competitions between the Alpha-agent and an ensemble of 19 agents were executed. Table 1 shows a list of agents that participated in the ensemble algorithm. The temperature parameters are $T = 0.7$ for soft-max and $\tau = 1$ for integration. Two yachts try to collect the garbage, which is generated at a new random position, faster than the other. They start at the same position, direction, and speed. We do not consider contact or interference between the two yachts in the simulator. Figure 5 shows a screenshot of the competitions.

When one yacht reaches the garbage position, a new garbage is set in a random position and the losing yacht comes to the same position, direction, and speed as the winner. If the two yachts reach the garbage at the same time-step, the result is a draw. One episode is completed when either of the yachts collects the garbage, and a positive reward of $+1000$ is given to the agents. In all the time-steps, a negative reward of -1 is given to all the agents.

We executed 1000 episodes. Table 2 shows the standings. A binominal test showed a significant difference ($p < 10^{-14}$).

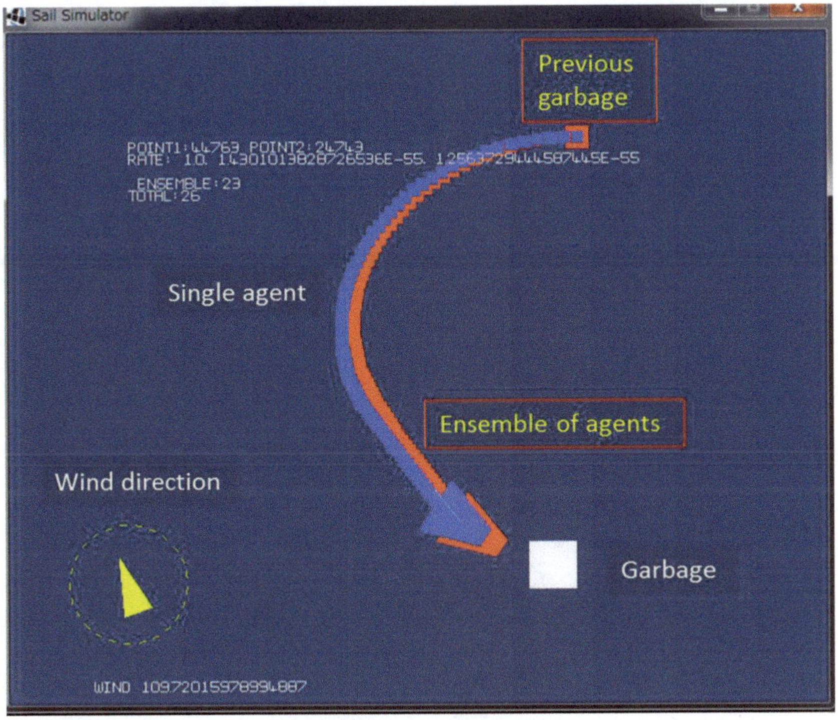

Fig. 5 Competitions between a single-agent and an ensemble of agents

Table 2 Standings of the competitions

	Wins	Losses	Draws
Ensemble	385	200	415
Single agent	200	385	415

5 Discussion

The results of experiment 2 show the effectiveness of the ensemble algorithm in Q-learning for garbage-collection sailing. Ensemble decision-making with diverse agents, state variables, learning parameters, and discretizing parameters worked well in the competitions. It is interesting that the participation of weak learners produces an advantage in decision making.

In this section, to clarify which agent contributed more to the winning episodes, we examine the sum of the Q-values of each agent for the selected action. Because the Q-value is back-propagated from a future reward, the larger Q-value an agent has, the more the agent experienced good decisions.

Table 3 shows the learning and discretizing parameters, state variables, and average Q-value for the selected actions by the agent in the winning episode. As in

Table 3 Parameters, state variables, and average Q-values

ID	α	γ	θ	State variables	Pre-learned	Average Q-value
2	0.01	0.99	45	s_1	No	913.404
3	0.01	0.99	–	s_2	No	1186.189
4	0.01	0.99	–	s_3	No	1198.595
5	0.01	0.99	45	s_1, s_2	No	1097.879
6	0.01	0.99	45	s_1, s_3	No	713.886
7	0.01	0.99	–	s_2, s_3	No	762.561
8	0.06	0.94	44	s_1	No	365.145
9	0.06	0.94	–	s_2	No	364.563
10	0.06	0.94	–	s_3	No	368.726
11	0.06	0.94	44	s_1, s_2	No	366.389
12	0.06	0.94	44	s_1, s_3	No	366.876
13	0.06	0.94	–	s_2, s_3	No	359.267
14	0.11	0.89	43	s_1	No	216.837
15	0.11	0.89	–	s_2	No	218.818
16	0.11	0.89	–	s_3	No	220.028
17	0.11	0.89	43	s_1, s_2	No	223.528
18	0.11	0.89	43	s_1, s_3	No	223.328
19	0.11	0.89	–	s_2, s_3	No	214.168

experiment 2, the ensemble-learning group's number of wins was statistically significantly greater than its number of losses in this experiment. The Alpha-agent, whose learning and discretizing parameters were $\alpha = 0.01$, $\gamma = 0.99$, and $\theta = 45$, was omitted from Table 3 because it was pre-learned and had greater Q-values than the others.

To summarize the results of this experiment:

- Agents with learning parameters similar to those of the Alpha-agent contribute more.
- Agents with a single state variable, especially s_2 or s_3, contribute more.
- Agents with two state variables, s_1 and s_2, also contribute to decision-making.

A larger learning rate and a smaller discount factor make an agent learn easily, but also forget easily. Thus, agents with IDs 8 to 19 tend to make more myopic decisions. State variable s_2 determines whether the yacht heads into the dead zone. The yacht loses speed in the dead zone. The agents who observe and learn s_2, e.g., agent ID 3, help avoid staying in the dead zone. State variable s_3 determines whether the target direction is in the dead zone. Agents who observe and learn s_3, e.g., agent ID 4, contribute to distinguish whether the yacht needs to zigzag. State variable s_1 is the target direction. Agent ID 5, who uses s_1 and s_2 as state variables, contributes more than Agent ID 2, who only uses s_1. This suggests the importance of wind direction in garbage-collection sailing.

6 Conclusion

A robot sailor can obtain its behaviour autonomously with reinforcement learning. However, reinforcement learning suffers from the curse of dimensionality, with an increase in state variables and an exponential increase in the number of states to realize fine control. This paper introduced an ensemble algorithm in Q-learning for garbage-collection sailing and discussed the effect. This paper especially investigated the enhancement of decision-making to sail faster to a target position, while keeping a small number of states and state variables. Numerical experiments showed a statistically significant enhancement from the proposed ensemble decision-making with diverse agents, state variables, and learning parameters.

References

1. Watkins, C.: Learning from Delayed Rewards Ph.D. thesis. Cambridge University, Cambridge (1989)
2. Sutton, R.S., Barto, A.G.: Reinforcement Learning: An Introduction. MIT Press, Cambridge (1998)
3. Barto, A.G., Sutton, R.S., Anderson, W.C.: Neuronlike adaptive elements that can solve difficult learning control problems. IEEE Trans. Syst., Man Cybern. **SMC–13**(5), 834–846 (1983)
4. Asada, M., Noda, S., Tawaratsumida, S., Hosoda, K.: Purposive behavior acquisition for a real robot by vision-based reinforcement learning. Mach. Learn. **23**, 279–303 (1996)
5. Notsu, A., Ichihashi, H., Honda, K.: State and action space segmentation algorithm in Q-learning. Int. Joint Conf. Neural Netw. pp. 2384–2389 (2008)
6. Kashimura, Y., Ueno, A., Tatsumi, S.: A continuous action space representation by particle filter for reinforcement learning. In: 22nd Annual Conference of the Japanese Society for Artificial Intelligence (2008) (in Japanese)
7. Morimoto, J., Doya, K.: Reinforcement learning of dynamic motor sequence: learning to stand up. In: Proceedings of IEEE/RSJ International Conference on Intelligent Robots and Systems, Vol. 3, 1721–1726 (1998)
8. Ishihara, S., Igarashi, H.: Applying the policy gradient method to behavior learning in multiagent systems: the pursuit problem. Syst. Comput. Jpn. **37**(10), 101–109 (2006)
9. Glorennec, P., Dept. d'Inf, I.d.R.F., Jouffe, L.: Fuzzy Q-learning. In: Proceeding of Sixth IEEE Internationall Conference on Fuzzy Systems, Vol. 2 (1997)
10. Hosoya, Y., Umano, M.: Dynamic fuzzy Q-learning with facility of tuning and removing fuzzy rules. IEEE World Congress on Computational Intelligence (2012)
11. Yairi, T., Hori, K., Nakasuka, S.: Autonomous reconstruction of state space for learning of robot behavior. In: Proceedings of of International Conference on Intelligent Robots and Systems, pp. 891–896 (2000)
12. Nagayoshi, M., Murao, H., Tamaki, H.: A reinforcement learning with switching controllers for a continuous action space. Artif. Life Robot. **15**, 97–100 (2010)
13. Sterne, P.J.: Reinforcement Sailing. Artificial Intelligence, School of Informatics, University of Edinburgh, Master of Science (2004)
14. Manabe, H., Tachibana, K.: Consideration of state representation for semi-autonomous reinforcement learning of sailing within a navigable area. Robot. Sail. **1**, 89–102 (2015)
15. Ueda, N., Nakano, R.: Generalization error of ensemble estimators. IEEE Conf. Neural Netw. pp. 90–95 (1996)
16. Sugiyama, T., Obata, T., Hoki, K., Ito, T.: Optimistic selection rule better than majority voting system. Comput. Games LNCS **6515**, 166–175 (2011)

17. Freund, Y., Schapire, E.R.: A decision-theoretic generalization of on-line learning and an application to boosting. J. Comput. Syst. Sci. **15**(1), 119–139 (1995)
18. Wiering, M.A., Hasselt, H.V.: Ensemble algorithm in reinforcement learning. IEEE Trans. Syst., Man Cybern. B **38**(4), 930–936 (2008)

Using a Controlled Sail and Tail to Steer an Autonomous Sailboat

Thomas Augenstein, Arjan Singh, Jesse Miller, Alex Pomerenk, Alec Dean and Andy Ruina

Abstract We are designing a cheap, mass producible 1 m semi-autonomous robotic sailboat that can navigate the oceans for months using only intermittent external supervision. The boat should efficiently collect environmental data such as salinity, turbidity, fluorescence, and animal sounds. The boat has a symmetric airfoil sail, a thin, bulbed keel, and most notably, an apparently new means of steering: a tail-vane rudder replacing the water rudder. The tail-vane rudder is to the main sail as an elevator is to the wing on an airplane, controlling both the angle-of-attack of the wing and the vehicle orientation. The angle-of-attack of the main wing is set by the tail-vane rudder, and the direction of the boat relative to the wind is set by the mast-rotation motor; the tail-vane rudder turns the boat. 2D and 3D dynamic simulations indicate that the tail-vane rudder design yields both (1) a stable wing angle-of-attack (like other wing-sail boats with hinged main sails); and also (2) boat directional stability relative to the wind (like boats with auto-steer wind-vanes). With fixed control-surface angles, the boat finds and maintains a stable heading, regardless of initial conditions. This directional stability allows the boat to operate intermittently with neither electrical power nor a complex wind-vane, thus reducing demands on batteries and solar cells. Tests show that in light winds the boat can sail stably within approximately ±45° of the wind direction. Because of the air tail-vane-, instead of water-, rudder, the boat requires new tacking techniques which we are currently

T. Augenstein (✉) · A. Singh · J. Miller · A. Pomerenk · A. Dean · A. Ruina
BioRobotics and Locomotion Laboratory, Cornell University, Ithaca, NY, USA
e-mail: tea43@cornell.edu

A. Singh
e-mail: as2995@cornell.edu

J. Miller
e-mail: ruina@cornell.edu

A. Pomerenk
e-mail: jam643@cornell.edu

A. Dean
e-mail: arp257@cornell.edu

A. Ruina
e-mail: agd57@cornell.edu

J.C. Alves and N.A. Cruz (eds.), *Robotic Sailing 2016*,
DOI 10.1007/978-3-319-45453-5_8

developing. As predicted by 3D simulations, experiments show that in heavy winds the boat has an oscillatory instability. It then finds a stable backwards sailing mode. We have yet-untested ideas for correcting this high-wind instability.

1 Motivation

Our boat aims to address deficiencies of other off-cost long-term environmental monitoring systems: manned vehicles cost on the order of \$10 K/day or more, satellites cannot collect in-situ data, buoys need to be deployed on site at a cost of tens of thousands of dollars per deployment, small gas or electric flying or motorized drones have limited range, and existing long distance drones such as the Saildrone [1], Wave Glider [2] and Seaglider [3] are relatively large (several meters) and cost tens of thousands of dollars each. On the other hand, our intended boats should be mass producible for less than \$10K each, have negligible operating costs, and collect off-coast data for years at a time.

The available-energy challenge. A small boat has limited access to electric power. The 3 kg of batteries that a 1 m boat might carry can store about 700 Wh of energy which, spread over a year, averages only 80 mW.

Solar cells do not much help this energy constraint: a 1 m boat has space for about 0.2 m^2 of solar panels, yielding a peak electrical harvest of up to 20 W at optimal conditions. More realistically, however, once we take into account the low angle of incidence at high latitudes (say, 60° North or South), short winter days, and cloud cover, the daily average solar harvest could average as low as 1/3 W (300 mW). Thus, our boat needs to receive navigation directions and GPS signals via satellite, read, process, and send sensor data, and operate its mechanical controls (sail trim and steering) on only 1/3 W average.

Our central challenge is this conflict between available and needed electrical power. The key to our approach to this challenge is intermittent control. The boat will alternate between two states: a "control-off" (idle) state that uses no electrical energy, and a "control-on" state, during which data collection and/or navigation mechanisms are active. The idle state is made possible by the boat's passive stability i.e. its ability to efficiently maintain a desired heading. This bi-modal operation, using a low duty factor (a low fraction of on-time), leads to low average energy use.

2 Two Kinds of Stability

During the idle control-off mode, the boat needs two kinds of stability: (1) the angle-of-attack of the sail must stably stay at near optimal (somewhere between the angles of attack for maximum lift and maximum lift-to-drag ratio); and (2) the boat heading must be kept near to the desired direction. That is, the boat needs both angle-of-attack stability and course stability (directional stability).

Conventional sailboats have no angle-of-attack stability, the sail angle is set by a human and does not respond to angle-of-attack. And most sailboats are, at best, only marginally directionally stable when sailing upwind, and not at all directionally stable when sailing downwind. To compensate for this marginal and/or absent stability, sailboats, like cars, are generally constantly steered, with the sail angle-of-attack being controlled by the sheets (the ropes which are used to change the sail angle in response to the sailor's sense of wind direction). For most robotic sailboats, the same approach is used, with robotic control replacing manual control of the sail and rudder angles.

However, both angle-of-attack stability and directional stability can be attained through passive mechanical means. Passive angle-of-attack stability can be achieved using a freely pivoting wing-sail with an attached downwind control surface (a vane-tail). Boats that have used this kind of angle-of-attack stability include Eldemer [4], SailDrone [1], the Walker Wingsail [5], the Zephyr, the Blue Nova [6], and Harbor Wing Technologies [7]. Additionally, this idea has been discussed by Elkaim [8] and Worsley [9]. However, such pivoting-sail angle-of-attack stability is essentially unrelated to directional (course) stability; all of the boats listed depend on more-or-less constantly-active rudders for course control.

Various mechanical and electrical contraptions yield course stability. For example, a wind-vane (e.g., [10]) can sense relative wind direction and, using a combination of wind and water power, then control the tiller/rudder motion to keep the boat's course constant relative to the relative-wind direction.

Another approach to (at least partly achieve) course stability has been proposed and tested by Sakurai et al. [11]. In a normal boat, heeling (leaning) moves the center of sail effort to leeward (downwind), thus causing a sail-generated torque on the boat to windward. Sakurai et al. reduces this problem by using a mast with fore-aft compliance. With increased wind force, the boat heels and the mast leans forward an almost equal angle. As a result, the line of force of the wind on the sail passes over the mast step. This net isotropic tipping of the mast (the net lean of the mast is in the direction of the wind force) effectively decouples the leaning from steering. Sakurai calls this effect directional stability. While this design does reduce the demand for rudder control by reducing unwanted steering due to wind fluctuations, the tipping mast does not stabilize the boat's course. Sakurai et al. attempts to achieve course stability by imitating the pitch stability of a glider. They think of the wing force on the keel and rudder as like the gravity force on the wing and horizontal stabilizer of an airplane. They carry out this reasoning-by-analogy using the reduced 2nd-order phugoid model [12]. They do not, however, model the full 2D dynamics (requiring 4th order ODEs) of a boat nor make an analogy with the full fourth-order model of airplane pitch stability (which does not have such a simple stability criterion). Nor do they note that the actual pitch angle of a plane, and thus the actual stable heading angle of a boat, would be controlled by the damping and not any simple combination of lift-surface angles. We are thus not fully convinced of the robustness of the course stability of their move-the-keel-back mechanism.

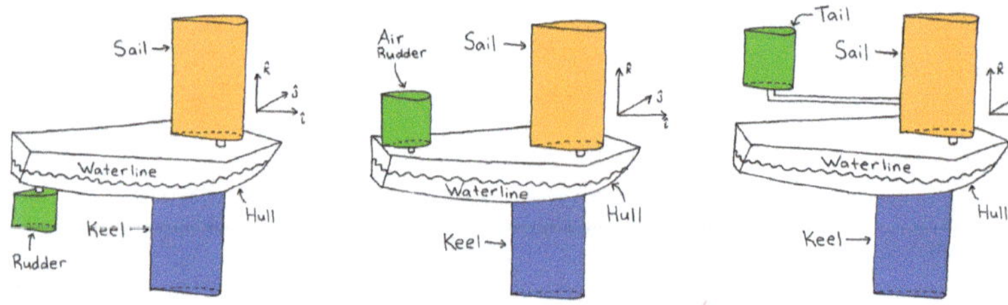

(a) Water Rudder (b) Air Rudder (c) Tail

Fig. 1 Three fixed-lift surface designs. Configuration (**c**), with a tail downwind of the wing, is our chosen design, called 'tail-vane'. Note that it has no water rudder

3 Our Approach to Stability

Our boat requires both kinds of stability; we seek the angle-of-attack stability that is achieved by a wing with a tail, and the directional stability that is achieved by a wind-vane contraption. However, we deem wind-vane mechanisms too delicate for months of untended exposure, and so seek a simpler approach. We seek a combination of lift surfaces that passively achieve both angle-of-attack and course stability, but without any moving parts.

A sailboat balances the forces on two wings: the sail above the water, and the keel below. In addition, it has an extra lift surface to control and/or stabilize the boat direction. In most boats this extra surface is a water rudder.

We considered three boat designs each of which uses three fixed-angle lift surfaces. We tested these in 2D simulations. All of the angles of the three lifting surfaces (wing, keel and control) were held rigid relative to the boat for the duration of every simulation. There was, for example, no hinge on the mast and no turning of the rudder. All boats had a wing and keel, vertically aligned. The designs were distinguished by the location of the third orientation-control surface. The three control-surface configurations were:

1. A water rudder i.e. a conventional sailboat, but with tiller position locked (Fig. 1a).
2. An air rudder (replacing the water rudder) at the stern of the boat (Fig. 1b).
3. A "tail-vane" air rudder (also replacing the water rudder), located downwind of the main wing. This configuration looks, above water, identical to the angle-of-attack vanes of, say, Saildrone. The two key differences between Saildrone's hinged sail-with-a-tail and our tail-vane rudder concept are (1) that our mast is not hinged, but has fixed angle relative to the boat between adjustments; and (2) our boat has no water rudder (Fig. 1c).

We hypothesized, and the 2D dynamical simulations confirmed, that of these three configurations, only the tail-vane air rudder would result in sailing with a stable angle-of-attack and a stable heading.

4 The Tail-Vane Air-Rudder

We chose the tail-vane air-rudder design and, for the reasons described above and below, we call our boat 'Sail-vane'. It is a sailing weather vane. Our choice configuration, the tail-vane air rudder (Fig. 1c), can be conceptualized a few ways, for example:

- **A weather vane with a boat underneath**. The sail and vane make up a weather vane. Because the main sail and keel are vertically aligned, together they make up an effective hinge. The tail-vane is turned slightly, misaligning the chord of the tail and main wing. Because only the tail has a moment about the keel, the tail always eventually points straight into the wind. This sets the angle-of-attack of the main wing. That is, the tail-vain chord angle is chosen to match the wing sail's desired angle-of-attack. Between passive control-off idle phases, the boat sets its heading by turning its mast-rotation motor, which rotates the boat beneath the sail.
- **A sideways airplane with a boat underneath**. As used by Lanchester in his famous phugoid plane models [12], the elevator of an airplane more-or-less sets the angle-of-attack of the main wing. Turn a rudderless airplane sideways (rotate 90° about the plane body axis) and you have the wing and air-rudder of our sail boat. The tail and sail are, together, like the wing and elevator of an airplane. As for other wing-sail boats with tails, our boat's wing-sail angle-of-attack is set by the tail angle. However, unlike other wing-sail boats, we also use the torque from the tail to stabilize the boat heading.

4.1 *More About Directional Stability*

A 3D, as opposed to 2D, dynamic simulation in MATLAB further tests the stability of the design. In these simulations we check:

1. Ability to sail in a desired direction regardless of the initial conditions of the boat (i.e. checking for 'global' attraction to a stable fixed point).
2. The range of sailing directions for which there are stable solutions.
3. The recovery to the desired direction after small disturbances (this covers, approximately, the linearized stability of a given sailing configuration).

Three dimensional testing was only performed on the tail-vane air-rudder configuration of Fig. 1c.

4.2 *Three Dimensional Simulations*

The dynamical simulations treat the boat as a three-dimensional rigid object subject to various fluid forces and gravity. The governing equations for the six-degree-of-freedom 3D boat were the six 2nd order differential equations of linear and angular

Table 1 Parameters used in our simulation to test the directional stability

True wind speed		5 m/s	Ballast mass		1.1 kg
Sail	Chord length	0.24 m	Water rudder	Chord Length	0.1 m
	Span	1 m		Span	0.1 m
	Mass	0.66 kg		Mass	0.4 kg
	Distance in front of C.O.M.	0 m		Distance in front of C.O.M.	–0.4 m
Keel	Chord length	0.04 m	Air rudder/Tail	Chord length	0.18 m
	Span	0.68 m		Span	0.46 m
	Mass	0.3 kg		Mass	0.4 kg
	Distance in front of C.O.M.	0 m		Distance in front of C.O.M.	–0.4 m

momentum balance. The forces on all lift surfaces were based on blade-element theory using published large-angle-of-attack lift and drag curves for a NACA 0015 wing. These forces took account of the angle of the wing and the relative fluid velocity due to boat motions and rotations. Buoyant forces and torques on the hull were based on an ellipsoidal model of the hull. Hull drag was based on scaling drag tests from a full sized boat.

This simulations use a 0.9 m long, 7 kg sailboat. A sample set of other parameters is in Table 1. Details of the simulations are in [13].

The two dimensional simulations used a similar approach, but neglected buoyancy forces and moments and constrained vertical motion of the boat, boat heeling, and boat pitching.

4.3 Simulation Results

The results of the simulations are expressed in Figs. 2b, 3, and 4 for the tail-vane, Fig. 5 for the water rudder, Fig. 6 for the air rudder. High wind issues that occur for the tail-vane are shown in Fig. 7.

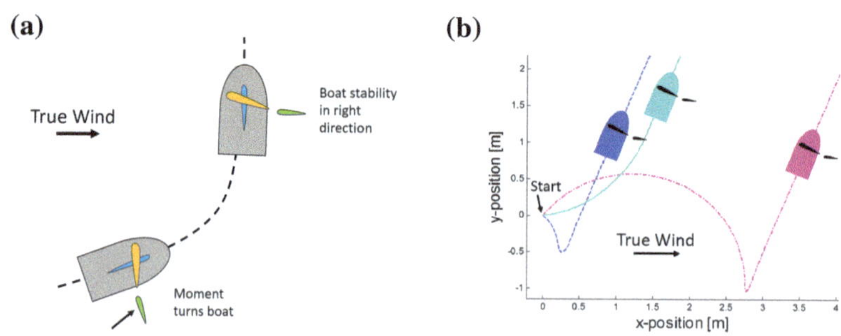

Fig. 2 Our sailboat uses a tail-vane control surface as a rudder with a controlled-angle sail. This configuration achieves directional stability

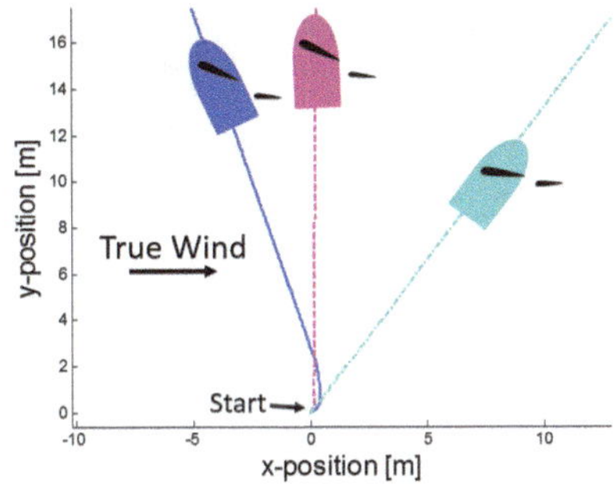

Fig. 3 2D simulated trajectories using three different tail-vane angles. All three simulations use the same initial conditions. The simulations show sailing upwind, cross wind and a broad reach. The tail-vane has a broad range of stable headings

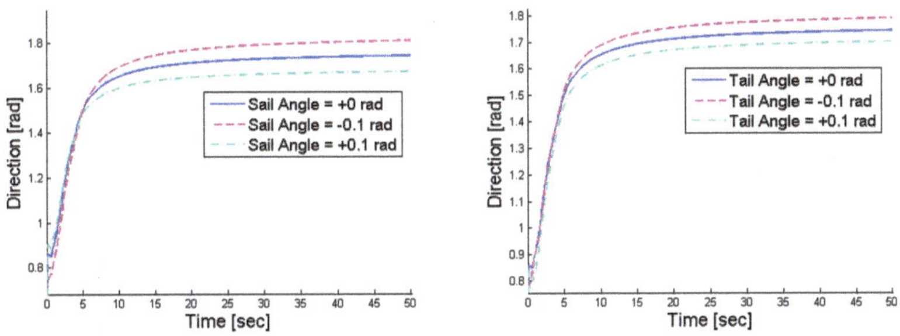

Fig. 4 Effect on the stable direction of the boat when changing the (*Left*) sail angle (*Right*) the tail-vane angle by ± 0.1 rad. There is a one-to-one ratio between the tail or sail angle and the angle of the boat's trajectory, indicating a high correlation between boat trajectory and small changes to the sail and tail-vane

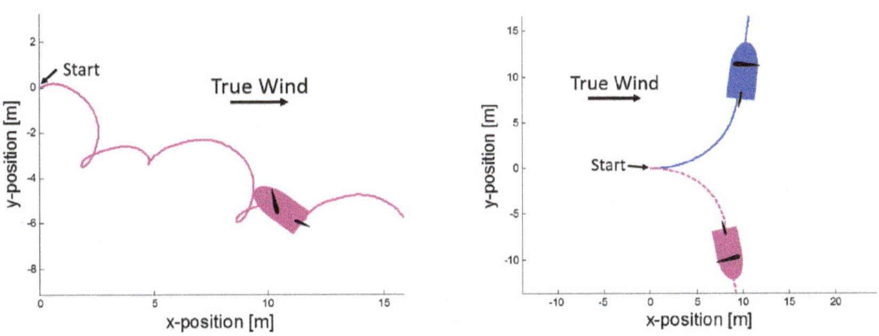

(a) Example of an unstable sailing trajectory caused by using a water rudder control surface.

(b) Sailboat trajectories produced by two separate trials using the water rudder.

Fig. 5 The water rudder failed both angle-of-attack or course stability tests

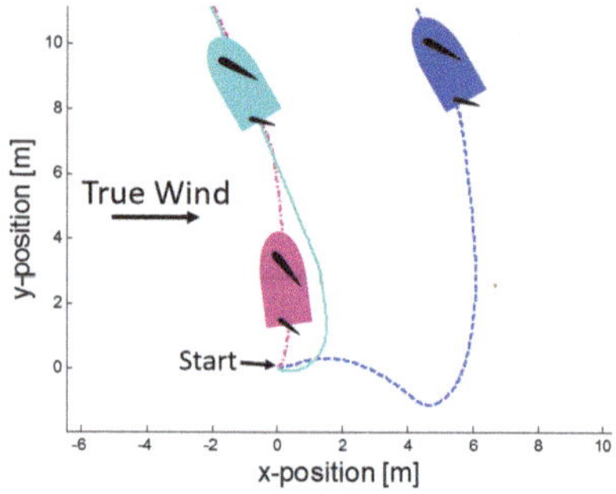

Fig. 6 Sailboat with an air rudder for three different initial orientations. Although the air rudder is course-stable on upwind trajectories, in high winds, with high heeling 3D simulations shows that it becomes unstable

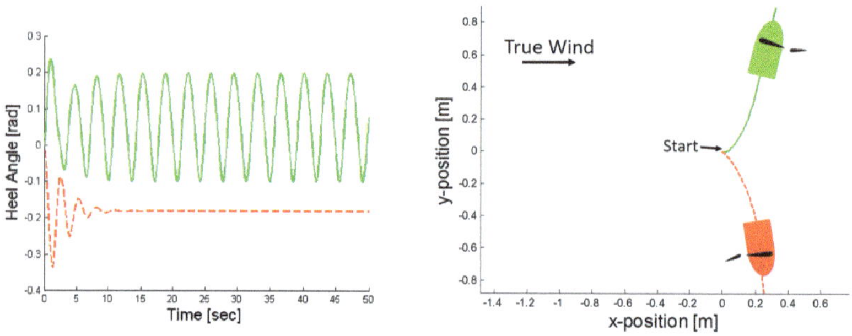

Fig. 7 (*Left*) Heeling angle of boat with tail-vane for two stable trajectories. (*Right*) Sailing trajectories for two trials using the tail-vane control surface. The boat has the same sail and tail angles for each trial but starts at a different orientation. High heeling can lead to dual stability, violating our directional stability criteria

4.3.1 Water Rudder Simulations

The actuated water rudder failed to meet any of the desired directional stability properties. Attempting to find a stable sailing direction typically resulted in an inefficient or unstable sailing trajectory. Figure 5a plots an example of this unstable trajectory. Conditions causing this trajectory were produced by setting the sail angle to 130° and the rudder angle to 10° relative to the hull, and held constant throughout the simulation.

Although most trajectories were unstable when using the water rudder, certain sail and rudder angles resulted in stable sailing directions. However, the stability of these directions was not robust: some headings close to the wind were unstable, headings more perpendicular to the wind were stable, and certain downwind headings exhibited bi-stability. For example, Fig. 5b illustrates a disturbance in the rudder angle of 6° causing nearly a 180° change in the stable sailing direction of the boat, indicating that the stable trajectory of the boat is overly sensitive to disturbances in the rudder angle. Due to this sensitivity, the water rudder fails to meet all of our desired stability criteria.

4.3.2 Air Rudder on the Hull

Simulations show that an air rudder on the hull, in the place of a conventional water rudder, leads to stable upwind sailing. The angle of the sail and rudder with respect to the boat were set to 40° and 60°, respectively. Three simulation trials were carried out, varying the initial boat heading in each. The trajectory of the boat, and the change in direction of this trajectory over time are illustrated in Fig. 6.

The air rudder was found to have a limited range of stable directions. In order for the sailboat using the air rudder to reach a stable direction, the resultant center of pressure on the sail and air rudder has to be downwind of the center of mass of the boat. In this case, a small disturbance in boat direction would result in an aerodynamic restoring moment on the sail/air rudder system. However, if the resultant center of pressure were in front of the center of mass, then a small disturbance causes the boat to become unstable. Therefore, depending on the positions of the sail and air rudder relative to the center of mass, the sailboat only has stable directions when sailing upwind or downwind, but not both. Due to the limited range of stable sailing directions, the air rudder fails to satisfy all of the desired characteristics of directional stability.

4.4 Tail-Vane Air Rudder Simulations

In 2D, the tail-vane control surface, combined with the actuated (but fixed between actuations) sail, were able to steer the boat control the angle of attack and be stable in idle control-off phases. Figure 2a illustrates this qualitatively.

Tests were conducted on the tail-vane control surface to confirm that each of the three directional stability criteria were met. In order to determine whether the sailboat would return to the same stable direction regardless of initial orientation, the simulation was run for three trials with varied initial boat headings. Figure 2b shows that the sailboat was, indeed, able to return to the desired heading regardless of its initial orientation.

A test was also performed to confirm that the sailboat can sail with directional stability both upwind and downwind. Figure 3 shows three separate trials run in the simulation where the sailboat was able to achieve trajectories upwind, on a broad reach, and perpendicular to the wind. The boat can be stable going directly downwind, but is more sensibly sailed on a broad reach, tacking to directly-down-wind destinations.

Simulations showed that a 0.1 rad change in sail and tail angles corresponds with about a 0.1 rad change in the stable direction of the boat, as illustrated in Fig. 4. This one-to-one correspondence between the boat parameters and angle of stability was expected, as the moment caused by the tail is the primary moment steering the boat.

4.5 *High-Wind 3D Issues*

In high winds, heeling is significant, causing a coupling of leaning to steering. The 3D simulations yielded the prediction that a given boat trim (given sail and tail angles) can correspond to multiple stable solutions. This is not a desirable result, since it indicates that the ultimate path of the boat depends on its initial condition. Figure 7 shows sailing trajectories for two separate trials with the sail and tail angles at 100° (relative to the boat's longitudinal axis from the bow) and 15° (relative to the sail), respectively. The green trajectory is the desired stable direction while the red trajectory is a newly-found 'parasitic' stable direction. The heel angle is also plotted for each trajectory. The heel angle shows persistent oscillation, probably indicating that the desired path undergoes a subcritical Hopf bifurcation with increasing wind speed. The undesired trajectory has a heel angle that decays to a constant value, indicating that the parasitic trajectory is more stable. The dual stability seems due to the coupling of lean with steer, and also with the nature of the post-stall lift and drag characteristics of the wing and keel.

On the intended boat trajectory, the center of pressure of the sail and tail system is downwind of the center of pressure of the underwater forces. Note that this solution is always predicted in the 2D simulations. Any disturbance from this equilibrium results in the tail exerting a restoring moment, giving the boat its directional stability.

In the unplanned configuration, shown by the red boat in Fig. 7, the tail is actually located upwind from the sail. This solution could not be stable if the boat was upright because a disturbance from the equilibrium would cause the tail to produce a destabilizing moment. However, because the boat is heeling, the force from the sail acts downwind of the underwater center of pressure. This indicates that it is possible for the stabilizing moment produced by the heeled sail to dominate the destabilizing moment from the tail (from this configuration, *de*stabilization is desirable).

5 Navigation Algorithm

Before worrying about, say, trans-Atlantic surveys, we have written a "short course" navigation algorithm. This takes as input a series of waypoints in global coordinates. These represent, more-or-less, buoys in a race course. The algorithm also takes as input the present boat position (from GPS), boat orientation (from a magnetic compass) and the wind direction (from a small wind-vane). The algorithm uses the boat polar (speed vs direction curves) to decide if a direct path to a target or an indirect path has the highest "velocity-made-good". The boat achieves the fastest heading by either heading directly towards the target, turning slightly, or by planning tacking or jibbing trajectories. While tacking works in remote control, jibbing with this design is more difficult (because, without slip rings, the mast cannot turn round and round).

6 Electrical Systems

We use an Arduino Due on a custom board to and from the Due. The sensor array is composed of a magnetic rotary encoder on a wind-vane, inertial measurement unit with compass (IMU), and a global positioning system sensor (GPS). The IMU measures the orientation of the boat, in terms of the roll, pitch and yaw. Our boat has two servomotors; one controls the sail (with respect to the hull) and one controls the tail-vane (with respect to the sail).

7 The Hull Shape

We chose a pre-existing, professionally-made racing hull shape, scaled the shape down to 1 m, and fabricated ourselves. This fast hull needed to support the 5 kg we expected our boat to weigh, meaning that its displacement specification must be greater than 5 kg after scaling the waterline length to 1 m to preserve race performance. We found that the *Luja*, a racing yacht by Sparkman and Stephens would nicely fit our purposes [14]. This hull is slightly wider than most racing hulls, giving it a larger displacement (larger volume to length-cubed ratio) but making it slightly slower. With a 1 m waterline length, *Luja* has a displacement of 6 kg, safely satisfying our 5 kg criteria. Using *Luja's* line plans (from a museum, scanned and put into CAD), we created a one-meter long foam plug, then a mold, and then fabricated from fiberglass and vinyl-ester with a gel coat finish. An image of the hull and the Sail-vane in action can be seen in Fig. 8.

8 Tests of the Physical Boat

As of this writing, the navigation algorithms have not been tested. But under radio control, the basic steering and passive stability have been tested on lake Cayuga in Ithaca. In light winds the boat steers, tacks and moves at a decent pace. In high winds i.e. high heeling, to our theoretical delight and practical concern, the oscillatory instability and bi-stable modes were observed. This is a victory for simulation and a practical challenge that needs to be overcome. These tests were composed of constant

(a) Bottom Surface of the Hull (b) Sailboat sailing to demonstrate low profile of hull

Fig. 8 The Sail-vane hull and the Sail-vane in action boat on the water

trajectory paths up, down, and across the wind and certain maneuvers including tacking, jibbing, and "directional stabilizing". Directional stabilizing means setting the sail and tail angle for the next desired heading abruptly and allowing the boat to fall into the heading. We performed these tests in both high wind (about 30 kmph) and low wind (about 10 kmph) environments. In the future, we hope to test our boat more thoroughly.

9 Conclusions and Future Work

The sail-vane concept, an air-rudder mounted downwind of the main-sale, shows promise for directional and angle-of-attack stable sailing. Thus there is promise for long term sailing with a low electrical-energy budget. However, as predicted by 3D simulation, the boat does not have the desired stability in high winds. We have several ideas for dealing with this problem, including:

1. A catamaran, trimaran or outrigger heel stabilization;
2. A deeper and/or heavier keel;
3. A moving ballast (the intermittent boat trimming would also trim the ballast location);
4. A fore-aft compliant mast, of the type used by Elkaim [8]. This would remove the coupling of learning, thus enabling the wind-vane stabilization to function despite high winds.
5. Using a smaller and/or shorter sail when high winds are possible.

We have not picked a strategy for choosing amongst these, all should work so it is a matter of trade-offs.

References

1. Saildrone Web Site. http://www.saildrone.com/
2. Waveglider Web Site. http://liquidr.com/
3. Seaglider. http://www.apl.washington.edu/projects/seaglider/summary.html
4. Udaya Shankar, N.: Reminiscing Rad. J. Astrophys. Astron. **32** 615 (2011). (Obituary of CV Radhakrishnan, see figure 4 on page 619 for picture of Radhakrishnan's wing-sail boat Eldemer in 2005)
5. Wingsail History—A Brief History of Wingsail Technology and its Commercial Applications. Wing Sails. http://www.solarnavigator.net/wing_sails.htm
6. Pando, J.: The Nacra Wingsail Project (2013). http://mywingsail.blogspot.com/
7. Harbor Wing Technologies Website. http://www.harborwingtech.com/index.htm
8. Elkaim, G.H.: Autonomous surface vehicle free-rotating wingsail section design and configuration analysis. J. Aircr. **45**(6), 1835–1852 (2008)
9. Worsley, P.: Peter Worsley—Wingsail Experiments—Article. http://www.sailwings.net/article.html
10. Windvane Self Steering Gear for Sailboats. Sailboat-Cruising.com. (2016). http://www.sailboat-cruising.com/self-steering-gear.html
11. Sakurai, A., Nakamura, T., Nakamoto, Y.: Experimental Study of a Directionally Stable Sailing Vehicle With a Free-Raking Rig and a Self-Trimming Sail. The 16th Chesapeake Sailing Yacht Symposium (2003)
12. Lanchester, F.W.: Aerodonetics: Constituting the Second Volume of a Complete Work on Aerial Flight. Van Nostrand (1909)
13. Miller, J.: A Directionally Stable Robotic Sailboat: Concept and Simulations, M-Eng thesis, Mechanical Engineering, Cornell University, April 2016. (preparation for the the Journal of Sailboat Technology)
14. Mystic Seaport Museum (2016) Luja Line Plans

FSC
www.fsc.org
MIX
Papier | Fördert gute Waldnutzung
FSC® C083411

Kolibri 360 GmbH
Ferdinand-Jühlke-Straße 7
99095 Erfurt, Deutschland
produktsicherheit@kolibri360.de